SpringerBriefs in Mathematics

SpringerBriefs in Mathematics showcases expositions in all areas of mathematics and applied mathematics. Manuscripts presenting new results or a single new result in a classical field, new field, or an emerging topic, applications, or bridges between new results and already published works, are encouraged. The series is intended for mathematicians and applied mathematicians.

More information about this series at http://www.springer.com/series/10030

Bicheng Yang · Michael Th. Rassias

On Hilbert-Type and Hardy-Type Integral Inequalities and Applications

Springer

Bicheng Yang
Department of Mathematics
Guangdong University of Education
Guangzhou, China

Michael Th. Rassias
Institute of Mathematics
University of Zurich
Zürich, Switzerland

Moscow Institute of Physics
and Technology
Dolgoprudny, Russia

Institute for Advanced Study
Program in Interdisciplinary Studies
Princeton, USA

ISSN 2191-8198 ISSN 2191-8201 (electronic)
SpringerBriefs in Mathematics
ISBN 978-3-030-29267-6 ISBN 978-3-030-29268-3 (eBook)
https://doi.org/10.1007/978-3-030-29268-3

Mathematics Subject Classification (2010): 26D15, 37A10, 47A07, 65B10

This Springer imprint is published by the registered company Springer Nature Switzerland AG
The registered company address is: Gewerbestrasse 11, 6330 Cham, Switzerland

As long as a branch of knowledge offers an abundance of problems, it is full of vitality.
David Hilbert

... we have always found with most inequalities, that we have a little new to add.

... in a subject (inequalities) like this, which has applications in every part of mathematics but never been developed systematically.
Godfrey H. Hardy

Preface

Hilbert-type inequalities including Hilbert's inequalities (originating in 1908), Hardy-Hilbert-type inequalities (originating in 1934) and Yang-Hilbert-type inequalities (originating in ca. 1998) have proven to be essential for various applications in mathematical analysis. These inequalities are mainly classified as integral inequalities, discrete inequalities and half-discrete inequalities. During the last two decades, research on these types of inequalities has flourished with the publication of several papers and books dedicated to this domain. A lot of activity has also been focused on Yang-Hilbert-type inequalities.

In the present monograph, by applying theories, methods and techniques of real analysis and functional analysis, we study four kinds of equivalent formulations of Hilbert-type inequalities, Hardy-type integral inequalities as well as their parametrized reverses. The best possible constant factors related mainly to the extended Hurwitz zeta function are presented in many examples. Furthermore, we consider the operator expressions with the norm and some particular analytic inequalities. Special cases of the above-mentioned four kinds of integral inequalities in the whole plane are considered. Through several lemmas and theorems, one can find within the present monograph an extensive account of these kinds of inequalities and operators.

The book comprises five chapters: In Chap. 1, we introduce some recent developments of Hilbert-type integral inequalities, discrete and half-discrete inequalities. In Chap. 2, by the application of theories, methods and techniques of real analysis, we provide several Hilbert-type integral inequalities with a general nonhomogeneous kernel. Moreover, the case of the general homogeneous kernel is also treated. In the form of applications, some special cases and a variety of examples—mainly related to the extended Hurwitz zeta function—are presented. In the last three chapters, we study further the cases of the reverse Hilbert-type integral inequalities, the Hardy-type integral inequalities and the reverse Hardy-type integral inequalities in the same spirit as in Chap. 2.

We hope that this monograph will prove to be useful especially to graduate students of mathematics, physics and engineering sciences.

Bicheng Yang
Department of Mathematics
Guangdong University of Education
Guangzhou, China

Michael Th. Rassias
Institute of Mathematics
University of Zurich
Zürich, Switzerland

Moscow Institute of Physics and Technology
Dolgoprudny, Russia

Institute for Advanced Study
Program in Interdisciplinary Studies
Princeton, USA

Contents

Chapter 1
Introduction

1.1 Hilbert's Inequalities and Their Operator Expressions

If $f(x), g(y) \geq 0$ $(x, y \in \mathbf{R}_+ = (0, \infty))$,

$$f, g \in L^2(\mathbf{R}_+) = \left\{ f; ||f||_2 = \left(\int_0^\infty |f(x)|^2 dx \right)^{\frac{1}{2}} < \infty \right\},$$

$$||f||_2, ||g||_2 > 0,$$

then we have the following, well known, Hilbert integral inequality as well as an equivalent form (cf. [1]):

$$\int_0^\infty \int_0^\infty \frac{f(x)g(y)}{x+y} dx dy < \pi ||f||_2 ||g||_2, \tag{1.1}$$

$$\left[\int_0^\infty \left(\int_0^\infty \frac{f(x)}{x+y} dx \right)^2 dy \right]^{\frac{1}{2}} < \pi ||f||_2, \tag{1.2}$$

where the constant factor π is the best possible.

If $a_m, b_n \geq 0$ $(m, n \in \mathbf{N} = \{1, 2, \ldots\})$,

$$a = \{a_m\}_{m=1}^\infty \in l^2 = \left\{ a; ||a||_2 = \left(\sum_{m=1}^\infty |a_m|^2 \right)^{\frac{1}{2}} < \infty \right\},$$

$$b = \{b_n\}_{n=1}^\infty \in l^2, \; ||a||_2, ||b||_2 > 0,$$

© The Author(s), under exclusive licence to Springer Nature Switzerland AG 2019
B. Yang and M. Th. Rassias, *On Hilbert-Type and Hardy-Type Integral
Inequalities and Applications*, SpringerBriefs in Mathematics,
https://doi.org/10.1007/978-3-030-29268-3_1

then we also have the following discrete Hilbert inequality and an equivalent form:

$$\sum_{n=1}^{\infty}\sum_{m=1}^{\infty}\frac{a_m b_n}{m+n} < \pi ||a||_2 ||b||_2, \tag{1.3}$$

$$\left[\sum_{n=1}^{\infty}\left(\sum_{m=1}^{\infty}\frac{a_m}{m+n}\right)^2\right]^{\frac{1}{2}} < \pi ||a||_2, \tag{1.4}$$

with the same best constant π (cf. [2]).

Using (1.2), we may define a Hilbert integral operator

$$T : L^2(\mathbf{R}_+) \to L^2(\mathbf{R}_+)$$

as follows (cf. [3]):

For any $f \in L^2(\mathbf{R}_+)$, there exists a $Tf \in L^2(\mathbf{R}_+)$, satisfying

$$Tf(y) = \int_0^{\infty}\frac{f(x)}{x+y}dx \, (y \in \mathbf{R}_+).$$

Then by (1.2), we have $||Tf||_2 < \pi ||f||_2$, and T is a bounded linear operator with

$$||T|| := \sup_{f(\neq \theta) \in L^2(\mathbf{R}_+)}\frac{||Tf||_2}{||f||_2} \leq \pi.$$

Since the constant factor in (1.2) is the best possible, we have $||T|| = \pi$.

Using (1.4), we may define the Hilbert operator $T : l^2 \to l^2$ as follows (cf. [4]):
For any $a = \{a_m\}_{m=1}^{\infty} \in l^2$, there exists a $Ta \in l^2$, satisfying

$$Ta(n) = \sum_{m=1}^{\infty}\frac{a_m}{m+n} \, (n \in \mathbf{N}).$$

Then by (1.4), we have $||Ta||_2 \leq \pi ||a||_2$, and T is a bounded linear operator with

$$||T|| := \sup_{a(\neq \theta) \in l^2}\frac{||Ta||_2}{||a||_2} \leq \pi.$$

Since the constant factor in (1.4) is the best possible, we have $||T|| = \pi$.

In 2002, Zhang [5] studied the above two operators and obtained some improvements of (1.1) and (1.3).

1.2   Two Classes of Hardy-Hilbert-Type Inequalities and Their Equivalent Forms

In 1925, by introducing one pair of conjugate exponents (p, q) $(\frac{1}{p} + \frac{1}{q} = 1)$, Hardy [6] proved the following extensions of (1.1)–(1.4):

Assuming that $p > 1$, $f(x), g(y) \geq 0$,

$$f \in L^p(\mathbf{R}_+) = \left\{ f; \|f\|_p = \left(\int_0^\infty |f(x)|^p dx \right)^{\frac{1}{p}} < \infty \right\},$$

$g \in L^q(\mathbf{R}_+)$, $\|f\|_p$, $\|g\|_q > 0$, we have the following Hardy-Hilbert integral inequality and the equivalent form:

$$\int_0^\infty \int_0^\infty \frac{f(x)g(y)}{x + y} dxdy < \frac{\pi}{\sin(\pi/p)} \|f\|_p \|g\|_q, \tag{1.5}$$

$$\left[\int_0^\infty \left(\int_0^\infty \frac{f(x)}{x + y} dx \right)^p dy \right]^{\frac{1}{p}} < \frac{\pi}{\sin(\pi/p)} \|f\|_p, \tag{1.6}$$

where the constant factor $\frac{\pi}{\sin(\pi/p)}$ is the best possible.

If $a_m, b_n \geq 0$,

$$a = \{a_m\}_{m=1}^\infty \in l^p = \left\{ a; \|a\|_p = \left(\sum_{m=1}^\infty |a_m|^p \right)^{\frac{1}{p}} < \infty \right\},$$

$$b = \{b_n\}_{n=1}^\infty \in l^q, \ \|a\|_p, \|b\|_q > 0,$$

then we also have the following equivalent discrete variant of the above inequalities:

$$\sum_{n=1}^\infty \sum_{m=1}^\infty \frac{a_m b_n}{m + n} < \frac{\pi}{\sin(\pi/p)} \|a\|_p \|b\|_q, \tag{1.7}$$

$$\left[\sum_{n=1}^\infty \left(\sum_{m=1}^\infty \frac{a_m}{m + n} \right)^p \right]^{\frac{1}{p}} < \frac{\pi}{\sin(\pi/p)} \|a\|_p, \tag{1.8}$$

with the same best possible constant factor $\frac{\pi}{\sin(\pi/p)}$.

For $p = q = 2$, the inequalities (1.5), (1.6), (1.7) and (1.8) reduce respectively to (1.1), (1.2), (1.3) and (1.4).

Definition 1.1 If $\lambda \in \mathbf{R} = (-\infty, \infty)$, $k_\lambda(x, y)$ is a measurable function in $\mathbf{R}_+^2 = \mathbf{R}_+ \times \mathbf{R}_+$, satisfying

$$k_\lambda(tx, ty) = t^{-\lambda}k_\lambda(x, y),$$

for any $t, x, y \in \mathbf{R}_+$, then we call $k_\lambda(x, y)$ a homogeneous function of degree $-\lambda$ in $\mathbf{R}_+^2$. If

$$k_\lambda(tm, tn) = t^{-\lambda}k_\lambda(m, n),$$

for any $t \in \mathbf{R}_+$, $m, n \in \mathbf{N}$, then we similarly call $k_\lambda(m, n)$ a homogeneous function of degree $-\lambda$ in $\mathbf{N}^2$.

In 1934, replacing $\frac{1}{x+y}$ by a general nonnegative homogeneous kernel of degree -1 such as $k_1(x, y)$, Hardy et al. [2] gave some extensions of (1.5), (1.6), (1.7) and (1.8) as follows:
Suppose that $p > 1$, $\frac{1}{p} + \frac{1}{q} = 1$,

$$k_p = \int_0^\infty k_1(u, 1)u^{\frac{-1}{p}} du \in \mathbf{R}_+.$$

If $f(x), g(y) \geq 0$, $f \in L^p(\mathbf{R}_+)$, $g \in L^q(\mathbf{R}_+)$, $||f||_p, ||g||_q > 0$, then we have the following Hardy-Hilbert-type integral inequality and the equivalent form:

$$\int_0^\infty \int_0^\infty k_1(x, y)f(x)g(y)dxdy < k_p||f||_p||g||_q, \tag{1.9}$$

$$\left[\int_0^\infty \left(\int_0^\infty k_1(x, y)f(x)dx\right)^p dy\right]^{\frac{1}{p}} < k_p||f||_p, \tag{1.10}$$

where the constant factor k_p is the best possible. If $k_1(m, n)m^{-\frac{1}{p}}$ ($k_1(m, n)n^{-\frac{1}{q}}$) is decreasing with respect to m $(n) \in \mathbf{N}$,

$$a_m, b_n \geq 0, a = \{a_m\}_{m=1}^\infty \in l^p, b = \{b_n\}_{n=1}^\infty \in l^q, ||a||_p, ||b||_q > 0,$$

then we also have the following equivalent discrete variant of the above inequalities:

$$\sum_{n=1}^\infty \sum_{m=1}^\infty k_1(m, n)a_mb_n < k_p||a||_p||b||_q, \tag{1.11}$$

$$\left[\sum_{n=1}^\infty \left(\sum_{m=1}^\infty k_1(m, n)a_m\right)^p\right]^{\frac{1}{p}} < k_p||a||_p, \tag{1.12}$$

with the same best possible constant factor k_p.

Some applications of Hardy-Hilbert-type inequalities are featured in [7].

1.3 Three Classes of Hilbert-Type Inequalities and Their Equivalent Forms

In 1998, by introducing an independent parameter $\lambda \in (0, \infty)$, Yang [8, 9] proved the following extension of (1.1):

For $f(x) \geq 0$ satisfying

$$0 < \int_0^\infty x^{1-\lambda} f^2(x) dx < \infty,$$

and $g(y) \geq 0$ satisfying

$$0 < \int_0^\infty y^{1-\lambda} g^2(y) dy < \infty,$$

we have

$$\int_0^\infty \int_0^\infty \frac{f(x)g(y)}{(x+y)^\lambda} dxdy < k_\lambda \left(\int_0^\infty x^{1-\lambda} f^2(x) dx \int_0^\infty y^{1-\lambda} g^2(y) dy \right)^{\frac{1}{2}},$$

(1.13)

where the constant factor $k_\lambda = B(\frac{\lambda}{2}, \frac{\lambda}{2})$ is the best possible and $B(u, v)$ $(u, v > 0)$ stands for the Beta function.

In 2004, by introducing two pairs of conjugate exponents (p, q), (r, s), namely

$$\frac{1}{p} + \frac{1}{q} = \frac{1}{r} + \frac{1}{s} = 1,$$

Yang [10] proved the following extension of (1.5):

For $p, r > 1$, $f(x) \geq 0$, satisfying

$$0 < \int_0^\infty x^{p(1-\frac{\lambda}{r})-1} f^p(x) dx < \infty,$$

and $g(y) \geq 0$, satisfying

$$0 < \int_0^\infty y^{q(1-\frac{\lambda}{s})-1} g^q(y) dy < \infty,$$

we have

$$\int_0^\infty \int_0^\infty \frac{f(x)g(y)}{x^\lambda + y^\lambda} dxdy$$

$$< \frac{\pi}{\lambda \sin(\pi/r)} \left[\int_0^\infty x^{p(1-\frac{\lambda}{r})-1} f^p(x) dx \right]^{\frac{1}{p}} \left[\int_0^\infty y^{q(1-\frac{\lambda}{s})-1} g^q(y) dy \right]^{\frac{1}{q}}, \qquad (1.14)$$

where the constant factor $\frac{\pi}{\lambda \sin(\pi/r)}$ is the best possible.

Recently, by introducing multi-parameters $\lambda_1, \lambda_2, \lambda \in \mathbf{R}$ ($\lambda_1 + \lambda_2 = \lambda$), Yang [11, 12] obtained some extensions of (1.9), (1.10), (1.11), (1.12) and (1.13) as follows:

Suppose that $k_\lambda(x, y)$ is a nonnegative homogeneous function of degree $-\lambda$, with

$$k(\lambda_1) := \int_0^\infty k_\lambda(t, 1)t^{\lambda_1 - 1}dt \in \mathbf{R}_+,$$

$$\phi(x) = x^{p(1-\lambda_1)-1}, \ \psi(y) = y^{q(1-\lambda_2)-1} \ (x, y \in \mathbf{R}_+).$$

If $f(x), g(y) \geq 0$,

$$f \in L_{p,\phi}(\mathbf{R}_+) = \left\{ f; \|f\|_{p,\phi} := \left(\int_0^\infty \phi(x)|f(x)|^p dx \right)^{\frac{1}{p}} < \infty \right\},$$

$$g \in L_{q,\psi}(\mathbf{R}_+), \|f\|_{p,\phi}, \|g\|_{q,\psi} > 0,$$

then we have the following Hilbert-type integral inequality and the equivalent form:

$$\int_0^\infty \int_0^\infty k_\lambda(x, y) f(x)g(y)dxdy < k(\lambda_1)\|f\|_{p,\phi}\|g\|_{q,\psi}, \qquad (1.15)$$

$$\left[\int_0^\infty \left(\int_0^\infty k_\lambda(x, y) f(x)dx \right)^p dy \right]^{\frac{1}{p}} < k(\lambda_1)\|f\|_{p,\phi}, \qquad (1.16)$$

where the constant factor $k(\lambda_1)$ is the best possible. Moreover, if $k_\lambda(x, y)$ attains a finite value and $k_\lambda(x, y)x^{\lambda_1 - 1}$ ($k_\lambda(x, y)y^{\lambda_2 - 1}$) is decreasing with respect to $x > 0$ ($y > 0$), then for $a_m, b_n \geq 0$,

$$a \in l_{p,\phi} = \left\{ a; \|a\|_{p,\phi} := (\sum_{m=1}^\infty \phi(m)|a_m|^p)^{\frac{1}{p}} < \infty \right\},$$

$$b = \{b_n\}_{n=1}^\infty \in l_{q,\psi}, \ \|a\|_{p,\phi}, \|b\|_{q,\psi} > 0,$$

we have the following equivalent discrete variant of the above inequalities:

$$\sum_{n=1}^\infty \sum_{m=1}^\infty k_\lambda(m, n)a_m b_n < k(\lambda_1)\|a\|_{p,\phi}\|b\|_{q,\psi}, \qquad (1.17)$$

$$\left[\sum_{n=1}^\infty \left(\sum_{m=1}^\infty k_\lambda(m, n)a_m \right)^p \right]^{\frac{1}{p}} < k(\lambda_1)\|a\|_{p,\phi}, \qquad (1.18)$$

where the constant factor $k(\lambda_1)$ is still the best possible.

Clearly, for $\lambda = 1$, $\lambda_1 = \frac{1}{q}$, $\lambda_2 = \frac{1}{p}$, (1.15), (1.16), (1.17) and (1.18) reduce respectively to (1.9), (1.10), (1.11) and (1.12); for $p = q = 2$, $\lambda_1 = \lambda_2 = \frac{\lambda}{2} > 0$, $k_\lambda(x, y) = \frac{1}{(x+y)^\lambda}$, (1.15) reduces to (1.13); for $\lambda_1 = \frac{\lambda}{r}$, $\lambda_2 = \frac{\lambda}{s}$, $k_\lambda(x, y) = \frac{1}{x^\lambda + y^\lambda}$ ($\lambda > 0$), (1.15) reduces to (1.14).

Regarding half-discrete Hilbert-type inequalities with nonhomogeneous kernels, Hardy et al. obtained a few results in Theorem 351 of [2]. However, they did not prove that the constant factors are the best possible. But, Yang [13] presented a result involving the kernel $\frac{1}{(1+nx)^\lambda}$ (by introducing a variable) and proved that the corresponding constant factor is the best possible.

Using weight functions and techniques of discrete and integral Hilbert-type inequalities with some additional conditions on the kernel, the following half-discrete Hilbert-type inequality and the equivalent forms with a general homogeneous kernel of degree $-\lambda \in \mathbf{R}$ and a best constant factor $k(\lambda_1)$ are obtained (cf. [14]):

$$\int_0^\infty f(x) \sum_{n=1}^\infty k_\lambda(x, n) a_n dx < k(\lambda_1) \|f\|_{p,\phi} \|a\|_{q,\psi}, \qquad (1.19)$$

$$\left[\sum_{n=1}^\infty n^{p\lambda_2 - 1} \left(\int_0^\infty k_\lambda(x, n) f(x) dx \right)^p \right]^{\frac{1}{p}} < k(\lambda_1) \|f\|_{p,\phi}, \qquad (1.20)$$

$$\left[\int_0^\infty x^{q\lambda_1 - 1} \left(\sum_{n=1}^\infty k_\lambda(x, n) a_n \right)^q dx \right]^{\frac{1}{q}} < k(\lambda_1) \|a\|_{q,\psi}. \qquad (1.21)$$

Additionally, a half-discrete Hilbert-type inequality with a general nonhomogeneous kernel $k_\lambda(1, xn)$ and a best constant factor is established by Yang [15]. Surveys on Hilbert-type inequalities are provided in [16, 17].

1.4   Some Results on Multidimensional Hilbert-Type Inequalities

If $i_0, j_0 \in \mathbf{N}$, $\alpha, \beta > 0$, we set

$$\|x\|_\alpha := \left(\sum_{k=1}^{i_0} |x_k|^\alpha \right)^{\frac{1}{\alpha}} \quad (x = (x_1, \ldots, x_{i_0}) \in \mathbf{R}^{i_0}),$$

$$\|y\|_\beta := \left(\sum_{k=1}^{j_0} |y_k|^\beta \right)^{\frac{1}{\beta}} \quad (y = (y_1, \ldots, y_{j_0}) \in \mathbf{R}^{j_0}).$$

In 2006, Hong [18] proved the following multidimensional Hilbert-type integral inequality and the equivalent form (for $\beta = \alpha$):

For $p > 1, \frac{1}{p} + \frac{1}{q} = 1, \lambda_1 + \lambda_2 = \lambda$,

$$\Phi(x) = x^{p(i_0-\lambda_1)-i_0} \ (x \in \mathbf{R}_+^{i_0}), \ \Psi(y) = y^{p(j_0-\lambda_2)-j_0} \ (y \in \mathbf{R}_+^{j_0}),$$

$$f(x) = f(x_1, \ldots, x_{i_0}) \geq 0, \ g(y) = g(y_1, \ldots, y_{j_0}) \geq 0,$$

$$0 < ||f||_{p,\Phi} = \left(\int_{\mathbf{R}_+^{i_0}} \Phi(x) f^p(x) dx \right)^{\frac{1}{p}} < \infty,$$

$$0 < ||g||_{q,\Psi} = \left(\int_{\mathbf{R}_+^{j_0}} \Psi(y) g^q(y) dy \right)^{\frac{1}{q}} < \infty,$$

we have the following equivalent inequalities with the kernel

$$k_\lambda(||x||_\alpha, ||y||_\alpha) = \frac{1}{(||x||_\alpha + ||y||_\alpha)^\lambda}$$

$(\lambda > 0)$:

$$\int_{\mathbf{R}_+^{j_0}} \int_{\mathbf{R}_+^{i_0}} k_\lambda(||x||_\alpha, ||y||_\alpha) f(x) g(y) dx dy < K(\lambda_1) ||f||_{p,\Phi} ||g||_{q,\Psi}, \tag{1.22}$$

$$\left[\int_{\mathbf{R}_+^{j_0}} ||y||_\alpha^{\frac{p\lambda}{s}-1} \left(\int_{\mathbf{R}_+^{i_0}} k_\lambda(||x||_\alpha, ||y||_\alpha) f(x) dx \right)^{\frac{1}{p}} dy \right]^{\frac{1}{p}} < K(\lambda_1) ||f||_{p,\Phi}, \tag{1.23}$$

where the particular constant factor

$$K(\lambda_1) = \frac{\Gamma^{i_0}(1/\alpha)}{\alpha^{i_0-1}\Gamma(i_0/\alpha)} B(\lambda_1, \lambda_2)$$

$(\lambda_1, \lambda_2 > 0)$ is the best possible.

In 2007, by introducing four particular kernels

$$k_\lambda(||x||_\alpha, ||y||_\alpha) = \frac{1}{|||x||_\alpha - ||y||_\alpha|^\lambda} \ (0 < \lambda < 1),$$

$$k_\lambda(||x||_\alpha, ||y||_\alpha) = \frac{\ln(||x||_\alpha/||y||_\alpha)}{||x||_\alpha^\lambda - ||y||_\alpha^\lambda} \ (\lambda > 0),$$

$$k_\lambda(||x||_\alpha, ||y||_\alpha) = \frac{1}{||x||_\alpha^\lambda + ||y||_\alpha^\lambda} \ (\lambda > 0)$$

and

$$k_\lambda(||x||_\alpha, ||y||_\alpha) = \frac{1}{(\max\{||x||_\alpha, ||y||_\alpha\})^\lambda} \quad (\lambda > 0),$$

Zhong et al. [19] established four pairs of equivalent inequalities (1.22) and (1.23) (for $\beta = \alpha$) with the best possible constant factors

$$K(\lambda_1) = \frac{\Gamma^{i_0}(1/\alpha)}{\alpha^{i_0-1}\Gamma(i_0/\alpha)} k(\lambda_1),$$

where

$$k(\lambda_1) = B(\lambda_1, 1 - \lambda) + B(\lambda_2, 1 - \lambda),$$

$$\left[\frac{\pi}{\lambda \sin(\pi\lambda_1/\lambda)}\right]^2, \quad \frac{\pi}{\lambda \sin(\pi\lambda_1/\lambda)}$$

and $\frac{\lambda}{\lambda_1\lambda_2}$ ($\lambda_1, \lambda_2 > 0$).

In 2011, Yang et al. [20] proved (1.22) and (1.23) with the general homogeneous kernel

$$k_\lambda(||x||_\alpha, ||y||_\beta)$$

and the best possible constant factor

$$K(\lambda_1) = \left(\frac{\Gamma^{j_0}(1/\beta)}{\beta^{j_0-1}\Gamma(j_0/\beta)}\right)^{\frac{1}{p}} \left(\frac{\Gamma^{i_0}(1/\alpha)}{\alpha^{i_0-1}\Gamma(i_0/\alpha)}\right)^{\frac{1}{q}} k(\lambda_1),$$

where,

$$k(\lambda_1) = \int_0^\infty k_\lambda(t, 1)t^{\lambda_1-1}dt \in \mathbf{R}_+.$$

In this case, for $i_0 = j_0 = \alpha = \beta$, (1.22) and (1.23) reduce respectively to (1.15) and (1.16).

In recent years, some results on multidimensional Hilbert-type integral inequalities are published in [21–25], and some results on multidimensional discrete as well as half-discrete Hilbert-type inequalities are presented in [26–40]. Other kinds of inequalities and operators are studied in [41–45] (see also [46–51]).

Remark 1.2 (1) Many different kinds of Hilbert-type discrete, half-discrete and integral inequalities, along with various applications have been presented during the last twenty years. Within this monograph we give special emphasis to certain original results that have been proved during the period 2009–2012. We present several generalizations, extensions and refinements of Hilbert-type discrete, half-discrete and integral inequalities involving many special functions such as the beta function, the gamma function, hypergeometric functions, trigonometric functions, hyperbolic functions, the Hurwitz zeta function, the Riemann zeta function, the Bernoulli functions, the Bernoulli numbers and Euler's constant (cf. [52–59]).

(2) In his six books, Yang [11, 12, 60–63] presented certain kinds of Hilbert-type operators with general homogeneous and nonhomogeneous kernels and two pairs of conjugate exponents as well as related inequalities. These research monographs contained recent developments of discrete, half-discrete and integral types of operators and inequalities along with their proofs, accompanied with examples and applications.

(3) In 2017, Hong [64] studied equivalent forms to a Hilbert-type integral inequality with a general homogeneous kernel with some parameters. Several authors continue this effort and have been investigating this topic for other types of integral inequalities and operators (cf. [59, 65–69]). Following this line of work, we consider equivalent properties to several kinds of Hilbert-type and Hardy-type integral inequalities as well as present applications involving the extended Hurwitz zeta function, the Riemann zeta function etc.

In Chap. 2 of the present book, by the use of weight functions and techniques of real analysis and functional analysis, we consider a few statements of Hilbert-type integral inequalities with a general nonhomogeneous kernel related to some parameters. The case of Hilbert-type integral inequalities with a general homogeneous kernel is deduced. Meanwhile, operator expressions and some examples mainly related to the Hurwitz zeta function are obtained in the form of applications. Additionally, a few statements of Hilbert-type integral inequalities in the whole plane with the exponent function as the integral variables are constructed. In the remaining three chapters, the cases of the reverse Hilbert-type integral inequalities, two kinds of Hardy-type integral inequalities as well as their reverses are considered similarly to Chap. 2.

References

1. Schur, I.: Bernerkungen sur theorie der beschrankten bilinearformen mit unendich vielen veranderlichen. J. Math. **140**, 1–28 (1911)
2. Hardy, G.H., Littlewood, J.E., Pólya, G.: Inequalities. Cambridge University Press, Cambridge (1934)
3. Carleman, I.: Sur les equations integrals singulieres a noyau reel et symetrique. Uppsala (1923)
4. Wilhelm, M.: On the spectrum of Hilbert's inequality. Am. J. Math. **72**, 699–704 (1950)
5. Zhang, K.W.: A bilinear inequality. J. Math. Anal. Appl. **271**, 288–296 (2002)
6. Hardy, G.H.: Note on a theorem of Hilbert concerning series of positive term. Proc. Lond. Math. Soc. **23**, 45–46 (1925)
7. Mitrinović, D.S., Pečarić, J.E., Fink, A.M.: Inequalities Involving Functions and Their Integrals and Derivatives. Kluwer Acaremic Publishers, Boston, USA (1991)
8. Yang, B.C.: On Hilbert's integral inequality. J. Math. Anal. Appl. **220**, 778–785 (1998)
9. Yang, B.C.: A note on Hilbert's integral inequality. Chin. Q. J. Math. **13**(4), 83–86 (1998)
10. Yang, B.C.: On an extension of Hilbert's integral inequality with some parameters. Aust. J. Math. Anal. Appl. **1**(1), Art.11: 1–8 (2004)
11. Yang, B.C.: Hilbert-Type Integral Inequalities. Bentham Science Publishers Ltd., The United Arab Emirates (2009)

12. Yang, B.C.: Discrete Hilbert-type Inequalities. Bentham Science Publishers Ltd., The United Arab Emirates (2011)
13. Yang, B.C.: A mixed Hilbert-type inequality with a best constant factor. Int. J. Pure Appl. Math. **20**(3), 319–328 (2005)
14. Yang, B.C., Chen, Q.: A half discrete Hilbert-type inequality with a homogeneous kernel and an extension. J. Ineq. Appl. **124** (2011)
15. Yang, B.C.: A half discrete Hilbert-type inequality with a non-homogeneous kernel and two variables. Mediterr. J. Math. **10**, 677–692 (2013)
16. Yang, B.C., Rassias, Th.M.: On the study of Hilbert-type inequalities with multi-parameters: a survey. Int. J. Nonlinear Anal. Appl. **2**(1), 21–34 (2011)
17. Debnath, L., Yang, B.C.: Recent developments of Hilbert-type discrete and integral inequalities with applications. Int. J. Math. Math. Sci. **2012**, Article ID 871845, 29 pages
18. Hong, H.: On multiple Hardy-Hilbert integral inequalities with some parameters. J. Inequal. Appl. **2006**, Article ID 94960, 11 pages
19. Zhong, W.Y., Yang B.C.: On a multiple Hilbert-type integral inequality with symmetric kernel. J. Inequal. Appl. **2007**, Article ID 27962, 17 pages
20. Yang, B.C., Krnić, M.: On the norm of a mult-dimensional Hilbert-type operator. Sarajevo J. Math. **7**(20), 223–243 (2011)
21. Krnić, M., Pećarić, J.E., Vuković, P.: On some higher-dimensional Hilbert's and Hardy-Hilbert's type integral inequalities with parameters. Math. Ineq. Appl. **11**, 701–716 (2008)
22. Yang, B.C.: Hilbert-type integral operators: norms and inequalities. In: Paralos, P.M., et al. (eds.) Nonlinear Analysis, Stability, Approximation, and Inequalities, pp. 771–859. Springer, New York (2012)
23. Rassias, M.Th., Yang B.C.: A multidimensional Hilbert-type integral inequality related to the Riemann zeta function. In: Daras, N.J. (eds.) Applications of Mathematics and Informatics in Science and Engineering, pp. 417–433. Springer, New York (2014)
24. Huang, Z.X., Yang, B.C.: A multidimensional Hilbert-type integral inequality. J. Inequal. Appl. **2015**, 151 (2015)
25. Liu, T., Yang, B.C., He, L.P.: On a multidimensional Hilbert-type integral inequality with logarithm function. Math. Inequal. Appl. **18**(4), 1219–1234 (2015)
26. Krnić, M., Vuković, P.: On a multidimensional version of the Hilbert-type inequality. Anal. Math. **38**, 291–303 (2012)
27. Yang, B.C., Chen, Q.: A multidimensional discrete Hilbert-type inequality. J. Math. Ineq. **8**(2), 267–277 (2014)
28. Chen, Q., Yang, B.C.: On a more accurate multidimensional Mulholland-type inequality. J. Ineq. Appl. **2014**, 322 (2014)
29. Yang, B.C.: Multidimensional discrete Hilbert-type inequalities, operator and compositions. In: Milovanović, G.V., et al. (eds.) Analytic Number Theory, Approximation Theory, and Special Functions, pp. 429–484. Springer, New York (2014)
30. Rassias, M.Th., Yang, B.C.: A multidimensional half-discrete Hilbert-type inequality and the Riemann zeta function. Appl. Math. Comput. **225**, 263–277 (2013)
31. Rassias, M.Th., Yang, B.C.: On a multidimensional half-discrete Hilbert-type inequality related to the hyperbolic cotangent function. Appl. Math. Comput. **242**, 800–813 (2014)
32. Yang, B.C.: On a more accurate multidimensional Hilbert-type inequality with parameters. Math. Inequal. Appl. **18**(2), 429–441 (2015)
33. Yang, B.C.: On a more accurate reverse multidimensional half-discrete Hilbert-type inequalities. Math. Inequal. Appl. **18**(2), 589–605 (2015)
34. Shi, Y.P., Yang, B.C.: On a multidimensional Hilbert-type inequality with parameters. J. Inequal. Appl. **2015**, 371 (2015)
35. Yang, B.C.: Multidimensional Hilbert-type integral inequalities and their operators expressions. In: Rassias, Th.M., Tóth, L. (eds.) Topics in Mathematical Analysis and Applications, pp. 769–814. Springer, New York (2015)
36. Yang, B.C.: Multidimensional half-discrete Hilbert-type inequalities and operator expressions. In: Rassias, Th.M., Pardalos, P.M. (eds.) Mathematics Without Boundaries, Surveys in Pure Mathematics, pp. 651–724. Springer, New York (2015)

37. Zhong, J.H., Yang, B.C.: An extension of a multidimensional Hilbert-type inequality. J. Inequal. Appl. **2017**, 78 (2017)
38. Yang, B.C., Chen, Q.: A more accurate multidimensional Hardy-Mulholland-type inequality with a general homogeneous kernel. J. Math. Inequal. **12**(1), 113–128 (2018)
39. Yang, B.C.: A more accurate multidimensional Hardy-Hilbert's inequality. J. Appl. Anal. Comput. **8**(2), 558–572 (2018)
40. Yang, B.C.: A more accurate multidimensional Hardy-Hilbert-type inequality. J. King Saud Univ. Sci. (2018). https://doi.org/10.1016/j.jksus.2018.01.004
41. Kato, T.: Perturbation theory for linear operators. Classics in Mathematics. Springer, Berlin (1995)
42. Kato, T.: Variation of discrete spectra. Comm. Math. Phys. **111**(3), 501–504 (1987)
43. Kato, T., Satake, I.: An algebraic theory of Landau-Kolmogorov inequalities. Tohoku Math. J. **33**(3), 421–428 (1981)
44. Kato, T.: On an inequality of Hardy, Littlewood, and Polya. Adv. Math. **7**, 217–218 (1971)
45. Kato, T.: Demicontinuity, hemicontinuity and monotonicity. II. Bull. Am. Math. Soc. **73**, 886–889 (1967)
46. Azar, L.E.: Two new forms of Hilbert integral inequality. Math. Ineq. Appl. **17**(3), 937–946 (2014)
47. Azar, L.E.: The connection between Hilbert and Hardy inequalities. J. Inequal. Appl. **2013**, 452. https://doi.org/10.1186/1029-242X-2013-452
48. Azar, L.E.: Some extensions of Hilbert integral inequality. J. Math. Ineq. **5**(1), 131–140 (2011)
49. Azar, L.E.: Some inequalities of Hilbert's type and applications. Jordan J. Math. Stat. **1**(2), 143–162 (2008)
50. Azar, L.E.: Two new forms of half-discrete Hilbert inequality. J. Egypt. Math. Soc. **22**(2), 254–257 (2014)
51. Azar, L.E.: On Some extensions of Hardy-Hilbert's inequality and applications. J. Inequal. Appl. **2008**, Article ID 546829, 14 pages. https://doi.org/10.1155/2008/546829
52. Rassias, M.Th., Yang, B.C.: A Hilbert-type integral inequality in the whole plane related to the hyper geometric function and the beta function. J. Math. Anal. Appl. **428**(2), 1286–1308 (2015)
53. Rassias, M.Th., Yang, B.C.: A more accurate half-discrete Hardy-Hilbert-type inequality with the best possible constant factor related to the extended Riemann-zeta function. Int. J. Nonlinear Anal. Appl. **7**(2), 1–27 (2016)
54. Rassias, M.Th., Yang, B.C.: A half-discrete Hilbert-type inequality in the whole plane related to the Riemann zeta function. Appl. Anal., https://doi.org/10.1080/00036811.2017.1313411
55. Rassias, M.Th., Yang, B.C.: Equivalent conditions of a Hardy-type integral inequality related to the extended Riemann zeta function. Adv. Oper. Theory **2**(3), 237–256 (2017)
56. Liao, J.Q., Yang, B.C.: On Hardy-type integral inequalities with the gamma function. J. Inequal. Appl. **2017**, 131 (2017)
57. Wang, A.Z., Yang, B.C.: A more accurate half-discrete Hardy-Hilbert-type inequality with the logarithmic function. J. Inequal. Appl. **2017**, 153 (2017)
58. Rassias, M.Th., Yang, B.C.: A Half-discrete Hardy-Hilbert-type inequality with a best possible constant factor related to the Hurwitz zeta function. In: Govil, N.K., Mohapatra, R.N., Qazi, M.A., Schmeisser, G. (eds.) Progression in Approximation Theory and Applicable Complex Analysis: In Memory of Q. I. Rahman, pp. 183–218. Springer, New York (2017)
59. Rassias, M.Th., Yang, B.C.: Equivalent properties of a Hilbert-type integral inequality with the best constant factor related the Hurwitz zeta function. Ann. Funct. Anal. **9**(2), 282–295 (2018)
60. Yang, B.C.: The Norm of Operator and Hilbert-Type Inequalities. Science Press, Beijing, China (2009)
61. Yang, B.C., Debnath, L.: Half-Discrete Hilbert-Type Inequalities. World Scientific Publishing Co. Pte. Ltd., Singapore (2014)
62. Yang, B.C.: Two Kinds of Multiple Half-Discrete Hilbert-Type Inequalities. Lambert Academic Publishing, Deutschland, Germany (2012)

63. Yang, B.C.: Topics on Half-Discrete Hilbert-Type Inequalities. Lambert Academic Publishing, Deutschland, Germany (2013)
64. Hong, Y.: On the structure character of Hilbert's type integral inequality with homogeneous kernel and applications. J. Jilin Univ. (Sci. Ed.) **55**(2), 189–194 (2017)
65. Hong, Y., Huang, Q.L., Yang, B.C., Liao, J.Q.: The necessary and sufficient conditions for the existence of a kind of Hilbert-type multiple integral inequality with the non-homogeneous kernel and its applications. J. Inequal. Appl. **2017**, 316 (2017)
66. Yang, B.C., Chen, Q.: Equivalent conditions of existence of a class of reverse Hardy-type integral inequalities with nonhomogeneous kernel. J. Jilin Univ. (Sci. Ed.) **55**(4), 804–808 (2017)
67. Yang, B.C.: Equivalent conditions of the existence of Hardy-type and Yang-Hilbert-type integral inequalities with the nonhomogeneous kernel. J. Guangdong Univ. Educ. **37**(3), 5–10 (2017)
68. Yang, B.C.: On some equivalent conditions related to the bounded property of Yang-Hilbert-type operator. J. Guangdong Univ. Educ. **37**(5), 5–11 (2017)
69. Yang, Z.M., Yang, B.C.: Equivalent conditions of the existence of the reverse Hardy-type integral inequalities with the nonhomogeneous kernel. J. Guangdong Univ. Educ. **37**(5), 28–32 (2017)

Chapter 2
Equivalent Statements of Hilbert-Type Integral Inequalities

In this chapter, by the use of methods of real analysis and weight functions, we consider a few equivalent statements of Hilbert-type integral inequalities with a general nonhomogeneous kernel related to certain parameters. In the form of applications, a few equivalent statements of Hilbert-type integral inequalities with a general homogeneous kernel are deduced. Moreover, we also consider operator expressions, a few particular cases, and some examples related to the extended Hurwitz zeta function as applications.

2.1 Two Lemmas

In the sequel of this chapter, we assume that

$$p > 1, \frac{1}{p} + \frac{1}{q} = 1, \sigma_1, \sigma, \mu \in \mathbf{R}, \sigma + \mu = \lambda,$$

$h(u)$ is a nonnegative measurable function in $\mathbf{R}_+$, such that

$$k(\sigma) := \int_0^\infty h(u)u^{\sigma-1}du \, (\geq 0). \tag{2.1}$$

For $n \in \mathbf{N}$, we define the following two expressions:

$$I_1 := \int_1^\infty \left(\int_0^1 h(xy)x^{\sigma+\frac{1}{pn}-1}dx \right) y^{\sigma_1-\frac{1}{qn}-1}dy, \tag{2.2}$$

$$I_2 := \int_0^1 \left(\int_1^\infty h(xy)x^{\sigma-\frac{1}{pn}-1}dx \right) y^{\sigma_1+\frac{1}{qn}-1}dy. \tag{2.3}$$

© The Author(s), under exclusive licence to Springer Nature Switzerland AG 2019
B. Yang and M. Th. Rassias, *On Hilbert-Type and Hardy-Type Integral Inequalities and Applications*, SpringerBriefs in Mathematics,
https://doi.org/10.1007/978-3-030-29268-3_2

Setting $u = xy$ in (2.2) and (2.3), by Fubini's theorem (cf. [1]), it follows that

$$
\begin{aligned}
I_1 &= \int_1^\infty \left[\int_0^y h(u) \left(\frac{u}{y} \right)^{\sigma + \frac{1}{pn} - 1} \frac{1}{y} du \right] y^{\sigma_1 - \frac{1}{qn} - 1} dy \\
&= \int_1^\infty y^{(\sigma_1 - \sigma) - \frac{1}{n} - 1} \left(\int_0^y h(u) u^{\sigma + \frac{1}{pn} - 1} du \right) dy \\
&= \int_1^\infty y^{(\sigma_1 - \sigma) - \frac{1}{n} - 1} dy \int_0^1 h(u) u^{\sigma + \frac{1}{pn} - 1} du \\
&\quad + \int_1^\infty y^{(\sigma_1 - \sigma) - \frac{1}{n} - 1} \int_1^y h(u) u^{\sigma + \frac{1}{pn} - 1} du \, dy \\
&= \int_1^\infty y^{(\sigma_1 - \sigma) - \frac{1}{n} - 1} dy \int_0^1 h(u) u^{\sigma + \frac{1}{pn} - 1} du \\
&\quad + \int_1^\infty \left[\int_u^\infty y^{(\sigma_1 - \sigma) - \frac{1}{n} - 1} dy \right] h(u) u^{\sigma + \frac{1}{pn} - 1} du, \quad\quad (2.4)
\end{aligned}
$$

$$
\begin{aligned}
I_2 &= \int_0^1 \left[\int_y^\infty h(u) \left(\frac{u}{y} \right)^{\sigma - \frac{1}{pn} - 1} \frac{1}{y} du \right] y^{\sigma_1 + \frac{1}{qn} - 1} dy \\
&= \int_0^1 y^{(\sigma_1 - \sigma) + \frac{1}{n} - 1} \left(\int_y^\infty h(u) u^{\sigma - \frac{1}{pn} - 1} du \right) dy \\
&= \int_0^1 y^{(\sigma_1 - \sigma) + \frac{1}{n} - 1} dy \int_y^1 h(u) u^{\sigma - \frac{1}{pn} - 1} du \\
&\quad + \int_0^1 y^{(\sigma_1 - \sigma) + \frac{1}{n} - 1} \int_1^\infty h(u) u^{\sigma - \frac{1}{pn} - 1} du \, dy \\
&= \int_0^1 \left[\int_0^u y^{(\sigma_1 - \sigma) + \frac{1}{n} - 1} dy \right] h(u) u^{\sigma - \frac{1}{pn} - 1} du \\
&\quad + \int_0^1 y^{(\sigma_1 - \sigma) + \frac{1}{n} - 1} dy \int_1^\infty h(u) u^{\sigma - \frac{1}{pn} - 1} du. \quad\quad (2.5)
\end{aligned}
$$

Lemma 2.1 *If $k(\sigma) > 0$, and if there exists a constant M, such that for any nonnegative measurable functions $f(x)$ and $g(y)$ in $(0, \infty)$, the following inequality*

$$
\begin{aligned}
I &:= \int_0^\infty \int_0^\infty h(xy) f(x) g(y) dx dy \\
&\leq M \left[\int_0^\infty x^{p(1-\sigma)-1} f^p(x) dx \right]^{\frac{1}{p}} \left[\int_0^\infty y^{q(1-\sigma_1)-1} g^q(y) dy \right]^{\frac{1}{q}} \quad\quad (2.6)
\end{aligned}
$$

holds true, then we have $\sigma_1 = \sigma$.

Proof Setting

$$k_1(\sigma) := \int_0^1 h(u)u^{\sigma-1}du$$

and

$$k_2(\sigma) := \int_1^\infty h(u)u^{\sigma-1}du,$$

it follows that

$$k(\sigma) = k_1(\sigma) + k_2(\sigma) > 0.$$

Since $k_i(\sigma) \geq 0$ $(i = 1, 2)$, without lose of generality, we assume that $k_1(\sigma) > 0$, namely, $h(u) > 0$ a.e. in an interval $I \subset (0, 1)$.

If $\sigma_1 < \sigma$, then for $n > \frac{1}{\sigma - \sigma_1}$ $(n \in \mathbf{N})$, we set two functions:

$$f_n(x) := \begin{cases} 0, 0 < x < 1 \\ x^{\sigma - \frac{1}{pn} - 1}, x \geq 1 \end{cases},$$

$$g_n(y) := \begin{cases} y^{\sigma_1 + \frac{1}{qn} - 1}, 0 < y \leq 1 \\ 0, y > 1 \end{cases}.$$

Hence, we obtain

$$J_2 := \left[\int_0^\infty x^{p(1-\sigma)-1} f_n^p(x)dx \right]^{\frac{1}{p}} \left[\int_0^\infty y^{q(1-\sigma_1)-1} g_n^q(y)dy \right]^{\frac{1}{q}}$$

$$= \left[\int_1^\infty x^{p(1-\sigma)-1} x^{p(\sigma - \frac{1}{pn} - 1)}dx \right]^{\frac{1}{p}} \left[\int_0^1 y^{q(1-\sigma_1)-1} y^{q(\sigma_1 + \frac{1}{qn} - 1)}dy \right]^{\frac{1}{q}}$$

$$= \left(\int_1^\infty x^{-\frac{1}{n}-1}dx \right)^{\frac{1}{p}} \left(\int_0^1 y^{\frac{1}{n}-1}dy \right)^{\frac{1}{q}} = n.$$

By (2.5), we have

$$\int_0^1 \left[\int_0^u y^{(\sigma_1-\sigma)+\frac{1}{n}-1}dy \right] h(u)u^{\sigma-1}du$$

$$\leq \int_0^1 \left[\int_0^u y^{(\sigma_1-\sigma)+\frac{1}{n}-1}dy \right] h(u)u^{\sigma-\frac{1}{pn}-1}du$$

$$\leq I_2 = \int_0^\infty \int_0^\infty h(xy) f_n(x)g_n(y)dxdy \leq MJ_2 = Mn < \infty. \qquad (2.7)$$

Since $(\sigma_1 - \sigma) + \frac{1}{n} < 0$, it follows that for any $u \in (0, 1)$,

$$\int_0^u y^{(\sigma_1-\sigma)+\frac{1}{n}-1}dy = \infty.$$

By (2.7), in view of the fact that $h(u)u^{\sigma-1} > 0$ a.e. in $I \subset (0, 1)$, we find that $\infty \leq Mn < \infty$, which is a contradiction.

If $\sigma_1 > \sigma$, then for $n > \frac{1}{\sigma_1-\sigma}$ ($n \in \mathbf{N}$), we set the following functions:

$$\tilde{f}_n(x) := \begin{cases} x^{\sigma+\frac{1}{pn}-1}, 0 < x \leq 1 \\ 0, x > 1 \end{cases},$$

$$\tilde{g}_n(y) := \begin{cases} 0, 0 < y < 1 \\ y^{\sigma_1-\frac{1}{qn}-1}, y \geq 1 \end{cases}.$$

Hence, we find

$$\tilde{J}_2 := \left[\int_0^\infty x^{p(1-\sigma)-1}\tilde{f}_n^p(x)dx\right]^{\frac{1}{p}}\left[\int_0^\infty y^{q(1-\sigma_1)-1}\tilde{g}_n^q(y)dy\right]^{\frac{1}{q}}$$

$$= \left[\int_0^1 x^{p(1-\sigma)-1}x^{p(\sigma+\frac{1}{pn}-1)}dx\right]^{\frac{1}{p}}\left[\int_1^\infty y^{q(1-\sigma_1)-1}y^{q(\sigma_1-\frac{1}{qn}-1)}dy\right]^{\frac{1}{q}}$$

$$= \left(\int_0^1 x^{\frac{1}{n}-1}dx\right)^{\frac{1}{p}}\left(\int_1^\infty y^{-\frac{1}{n}-1}dy\right)^{\frac{1}{q}} = n.$$

By (2.4), we have

$$\int_1^\infty y^{(\sigma_1-\sigma)-\frac{1}{n}-1}dy \int_0^1 h(u)u^{\sigma+\frac{1}{pn}-1}du$$

$$\leq I_1 = \int_0^\infty \int_0^\infty h(xy)\tilde{f}_n(x)\tilde{g}_n(y)dxdy \leq M\tilde{J}_2 = Mn < \infty. \qquad (2.8)$$

Since $(\sigma_1 - \sigma) - \frac{1}{n} > 0$, it follows that

$$\int_1^\infty y^{(\sigma_1-\sigma)-\frac{1}{n}-1}dy = \infty.$$

By (2.8), in view of

$$\int_0^1 h(u)u^{\sigma+\frac{1}{pn}-1}du > 0,$$

we have $\infty \leq Mn < \infty$, which is a contradiction.

Hence, we conclude that $\sigma_1 = \sigma$.

This completes the proof of the lemma. $\qquad \square$

For $\sigma_1 = \sigma$, we still have

Lemma 2.2 *There exists a constant M, such that if for any nonnegative measurable functions $f(x)$ and $g(y)$ in $(0, \infty)$ the following inequality*

$$\int_0^\infty \int_0^\infty h(xy)f(x)g(y)dxdy$$

$$\leq M \left[\int_0^\infty x^{p(1-\sigma)-1} f^p(x)dx \right]^{\frac{1}{p}} \left[\int_0^\infty y^{q(1-\sigma)-1} g^q(y)dy \right]^{\frac{1}{q}} \qquad (2.9)$$

holds true, then we have $k(\sigma) \leq M < \infty$.

Proof For $\sigma_1 = \sigma$, we reduce the inequality to (2.4) and then use inequality $I_1 \leq M \tilde{J}_2$ (when $\sigma_1 = \sigma$) as follows:

$$\frac{1}{n} I_1 = \frac{1}{n} \left[\int_1^\infty y^{-\frac{1}{n}-1} dy \int_0^1 h(u)u^{\sigma+\frac{1}{pn}-1} du \right.$$

$$\left. + \int_1^\infty \left(\int_u^\infty y^{-\frac{1}{n}-1} dy \right) h(u)u^{\sigma+\frac{1}{pn}-1} du \right]$$

$$= \int_0^1 h(u)u^{\sigma+\frac{1}{pn}-1} du + \int_1^\infty h(u)u^{\sigma-\frac{1}{qn}-1} du$$

$$\leq \frac{1}{n} M \tilde{J}_2 = M. \qquad (2.10)$$

By Fatou's lemma (cf. [1]) and (2.10), we have

$$k(\sigma) = \int_0^1 \lim_{n \to \infty} h(u)u^{\sigma+\frac{1}{pn}-1} du + \int_1^\infty \lim_{n \to \infty} h(u)u^{\sigma-\frac{1}{qn}-1} du$$

$$\leq \underline{\lim}_{n \to \infty} \left[\int_0^1 h(u)u^{\sigma+\frac{1}{pn}-1} du + \int_1^\infty h(u)u^{\sigma-\frac{1}{qn}-1} du \right]$$

$$\leq M \; (< \infty). \qquad (2.11)$$

This completes the proof of the lemma.

2.2 Main Results and Some Corollaries

Theorem 2.3 *Suppose that M is a constant. The following Statements (i), (ii) and (iii) are equivalent:*

(i) For any nonnegative measurable function $f(x)$ in $(0, \infty)$, we have the following inequality:

$$J := \left[\int_0^\infty y^{p\sigma_1 - 1} \left(\int_0^\infty h(xy)f(x)dx \right)^p dy \right]^{\frac{1}{p}}$$

$$\leq M \left[\int_0^\infty x^{p(1-\sigma)-1} f^p(x)dx \right]^{\frac{1}{p}}. \tag{2.12}$$

(ii) *For any nonnegative measurable functions* $f(x)$ *and* $g(y)$ *in* $(0, \infty)$, *we have the following inequality:*

$$I = \int_0^\infty \int_0^\infty h(xy)f(x)g(y)dxdy$$

$$\leq M \left[\int_0^\infty x^{p(1-\sigma)-1} f^p(x)dx \right]^{\frac{1}{p}} \left[\int_0^\infty y^{q(1-\sigma_1)-1} g^q(y)dy \right]^{\frac{1}{q}}. \tag{2.13}$$

(iii) *For* $k(\sigma) > 0$, *we have* $\sigma_1 = \sigma$, *and* $k(\sigma) \leq M \ (< \infty)$.

Proof $(i) \Rightarrow (ii)$. By Hölder's inequality (cf. [2]), we find

$$I = \int_0^\infty \left(y^{\sigma_1 - \frac{1}{p}} \int_0^\infty h(xy)f(x)dx \right) \left(y^{\frac{1}{p} - \sigma_1} g(y) \right) dy$$

$$\leq J \left[\int_0^\infty y^{q(1-\sigma_1)-1} g^q(y)dy \right]^{\frac{1}{q}}. \tag{2.14}$$

Then by (2.12), we have (2.13).

$(ii) \Rightarrow (iii)$. Since $k(\sigma) > 0$, by Lemma 2.1, we have $\sigma_1 = \sigma$. Then by Lemma 2.2, we have $k(\sigma) \leq M \ (< \infty)$.

$(iii) \Rightarrow (i)$. For fixed $y > 0$, setting $u = xy$, we obtain the following weight function:

$$\omega(\sigma, y) := y^\sigma \int_0^\infty h(xy)x^{\sigma-1}dx = \int_0^\infty h(u)u^{\sigma-1}du = k(\sigma) \ (y \in \mathbf{R}_+). \tag{2.15}$$

By Hölder's inequality with weight and (2.15), we have

$$\left(\int_0^\infty h(xy)f(x)dx \right)^p$$

$$= \left\{ \int_0^\infty h(xy) \left[\frac{y^{(\sigma-1)/p}}{x^{(\sigma-1)/q}} f(x) \right] \left[\frac{x^{(\sigma-1)/q}}{y^{(\sigma-1)/p}} \right] dx \right\}^p$$

$$\leq \int_0^\infty h(xy) \frac{y^{\sigma-1}}{x^{(\sigma-1)p/q}} f^p(x)dx \left[\int_0^\infty h(xy) \frac{x^{\sigma-1}}{y^{(\sigma-1)q/p}} dx \right]^{p/q}$$

$$= \left[\omega(\sigma, y)y^{q(1-\sigma)-1}\right]^{p-1} \int_0^\infty h(xy)\frac{y^{\sigma-1}}{x^{(\sigma-1)p/q}} f^p(x)dx$$

$$= (k(\sigma))^{p-1}y^{-p\sigma+1} \int_0^\infty h(xy)\frac{y^{\sigma-1}}{x^{(\sigma-1)p/q}} f^p(x)dx. \qquad (2.16)$$

For $\sigma_1 = \sigma$, by Fubini's theorem (cf. [1]) and (2.16), we obtain

$$J \le (k(\sigma))^{\frac{1}{q}} \left[\int_0^\infty \int_0^\infty h(xy)\frac{y^{\sigma-1}}{x^{(\sigma-1)p/q}} f^p(x)dxdy\right]^{\frac{1}{p}}$$

$$= (k(\sigma))^{\frac{1}{q}} \left\{\int_0^\infty \left[\int_0^\infty h(xy)\frac{y^{\sigma-1}}{x^{(\sigma-1)(p-1)}}dy\right] f^p(x)dx\right\}^{\frac{1}{p}}$$

$$= (k(\sigma))^{\frac{1}{q}} \left[\int_0^\infty \omega(\sigma, x)x^{p(1-\sigma)-1} f^p(x)dx\right]^{\frac{1}{p}}$$

$$= k(\sigma) \left[\int_0^\infty x^{p(1-\sigma)-1} f^p(x)dx\right]^{\frac{1}{p}}.$$

For $k(\sigma) \le M$, (2.12) follows.

Therefore, the Statements (i), (ii) and (iii) are equivalent.

This completes the proof of the theorem.

For $\sigma_1 = \sigma$, by Theorem 2.3 and Lemma 2.2, we have the following:

Theorem 2.4 *Suppose that M is a constant. The following Statements (i), (ii) and (iii) are equivalent:*

(i) For any nonnegative measurable function $f(x)$ in $(0, \infty)$, we have the following inequality:

$$J_1 := \left[\int_0^\infty y^{p\sigma-1} \left(\int_0^\infty h(xy)f(x)dx\right)^p dy\right]^{\frac{1}{p}}$$

$$\le M \left[\int_0^\infty x^{p(1-\sigma)-1} f^p(x)dx\right]^{\frac{1}{p}}. \qquad (2.17)$$

(ii) For any nonnegative measurable functions $f(x)$ and $g(y)$ in $(0, \infty)$, we have the following inequality:

$$I = \int_0^\infty \int_0^\infty h(xy)f(x)g(y)dxdy$$

$$\le M \left[\int_0^\infty x^{p(1-\sigma)-1} f^p(x)dx\right]^{\frac{1}{p}} \left[\int_0^\infty y^{q(1-\sigma)-1} g^q(y)dy\right]^{\frac{1}{q}}. \qquad (2.18)$$

(iii) $k(\sigma) \leq M < \infty$.
Moreover, if (iii) holds true, then the constant factor $M = k(\sigma)$ in (2.17) and (2.18) is the best possible.

Proof $(i) \Rightarrow (ii)$. If (2.17) holds true, then by (2.14) (for $\sigma_1 = \sigma$), we derive (2.18).
$(ii) \Rightarrow (iii)$. By Lemma 2.2, we have that

$$k(\sigma) \leq M < \infty.$$

$(iii) \Rightarrow (i)$. By the proof of $(iii) \Rightarrow (i)$ in Theorem 2.3, we obtain (2.17).
Therefore, the Statements (i), (ii) and (iii) are equivalent.
If Statement (iii) holds true, and there exists a constant $M \leq k(\sigma)$, such that (2.18) is valid, then by (iii), we have $k(\sigma) \leq M$. It follows that $M = k(\sigma)$ is the best possible constant factor of (2.18). If the constant factor $M = k(\sigma)$ in (2.17) is not the best possible, then by (2.14) (for $\sigma_1 = \sigma$) we would reach a contradiction that the constant factor in (2.18) is not the best possible.
This completes the proof of the theorem.   □

For $M = k(\sigma)$, we have

Theorem 2.5 *Suppose that $k(\sigma) \in \mathbf{R}_+$. The following Statements (i) and (ii) are valid and equivalent:*
(i) For any $f(x) \geq 0$, satisfying

$$0 < \int_0^\infty x^{p(1-\sigma)-1} f^p(x)dx < \infty,$$

we have the following inequality:

$$J_1 = \left[\int_0^\infty y^{p\sigma-1} \left(\int_0^\infty h(xy)f(x)dx \right)^p dy \right]^{\frac{1}{p}}$$
$$< k(\sigma) \left[\int_0^\infty x^{p(1-\sigma)-1} f^p(x)dx \right]^{\frac{1}{p}}. \tag{2.19}$$

(ii) For any $f(x) \geq 0$, satisfying

$$0 < \int_0^\infty x^{p(1-\sigma)-1} f^p(x)dx < \infty,$$

and $g(y) \geq 0$, satisfying

$$0 < \int_0^\infty y^{q(1-\sigma)-1} g^q(y)dy < \infty,$$

we have the following inequality:

$$I = \int_0^\infty \int_0^\infty h(xy) f(x) g(y) dx dy$$

$$< k(\sigma) \left[\int_0^\infty x^{p(1-\sigma)-1} f^p(x) dx \right]^{\frac{1}{p}} \left[\int_0^\infty y^{q(1-\sigma)-1} g^q(y) dy \right]^{\frac{1}{q}}. \quad (2.20)$$

Moreover, the constant factor $k(\sigma)$ in (2.19) and (2.20) is the best possible.
In particular, for $\sigma = \frac{1}{p}$, we have the following equivalent inequality with the best possible constant factor

$$k(\frac{1}{p}) = \int_0^\infty h(u) u^{-\frac{1}{q}} du \in \mathbf{R}_+ :$$

$$\left[\int_0^\infty \left(\int_0^\infty h(xy) f(x) dx \right)^p dy \right]^{\frac{1}{p}} < k\left(\frac{1}{p}\right) \left(\int_0^\infty x^{p-2} f^p(x) dx \right)^{\frac{1}{p}}, \quad (2.21)$$

$$\int_0^\infty \int_0^\infty h(xy) f(x) g(y) dx dy$$

$$< k\left(\frac{1}{p}\right) \left(\int_0^\infty x^{p-2} f^p(x) dx \right)^{\frac{1}{p}} \left(\int_0^\infty g^q(y) dy \right)^{\frac{1}{q}}. \quad (2.22)$$

Proof We first prove that (2.19) is valid. In the proof of $(iii) \Rightarrow (i)$ in Theorem 2.3, if (2.16) assumes the form of equality for some $y \in (0, \infty)$, then (cf. [2]) there exist constants A and B, such that they are not all zero, and

$$A \frac{y^{\sigma-1}}{x^{(\sigma-1)p/q}} f^p(x) = B \frac{x^{\sigma-1}}{y^{(\sigma-1)q/p}} \quad a.e. \text{ in } \mathbf{R}_+.$$

Let us suppose that $A \neq 0$ (otherwise $B = A = 0$). Then it follows that

$$x^{p(1-\sigma)-1} f^p(x) = y^{q(1-\sigma)} \frac{B}{Ax} \quad a.e. \text{ in } \mathbf{R}_+,$$

which contradicts the fact that

$$0 < \int_0^\infty x^{p(1-\sigma)-1} f^p(x) dx < \infty.$$

Hence, (2.16) takes the form of strict inequality. By the last proof of $(iii) \Rightarrow (i)$ in Theorem 2.20 (for $\sigma_1 = \sigma$), we have (2.19).

$(ii) \Rightarrow (i)$. By (2.14) (for $\sigma_1 = \sigma$), we obtain (2.20).

$(i) \Rightarrow (ii)$. We set the following function:

$$g(y) := y^{p\sigma-1} \left(\int_0^\infty h(xy) f(x) dx \right)^{p-1} \quad (y > 0).$$

If $J_1 = \infty$, we get a contradiction since the right hand side of (2.19) is finite; if $J_1 = 0$, then (2.19) is trivially valid. We suppose that $0 < J_1 < \infty$. By (2.20), we get that

$$0 < \int_0^\infty y^{q(1-\sigma)-1} g^q(y) dy = J_1^p = I$$

$$< k(\sigma) \left[\int_0^\infty x^{p(1-\sigma)-1} f^p(x) dx \right]^{\frac{1}{p}} \left[\int_0^\infty y^{q(1-\sigma)-1} g^q(y) dy \right]^{\frac{1}{q}} < \infty,$$

$$J_1 = \left[\int_0^\infty y^{q(1-\sigma)-1} g^q(y) dy \right]^{\frac{1}{p}} < k(\sigma) \left[\int_0^\infty x^{p(1-\sigma)-1} f^p(x) dx \right]^{\frac{1}{p}},$$

namely, (2.19) holds true.

Therefore, the statements (i) and (ii) are valid and equivalent.

Moreover, by Theorem 2.4, the constant factor $k(\sigma)$ in (2.19) and (2.20) is the best possible.

This completes the proof of the theorem.   □

Setting

$$y = \frac{1}{Y}, \quad G(Y) = g\left(\frac{1}{Y}\right) \frac{1}{Y^2}$$

in Theorems 2.3, 2.4 and 2.5, then replacing Y by y, we obtain the following corollary:

Corollary 2.6 *Suppose that M is a constant. The following Statements (i), (ii) and (iii) are equivalent:*

(i) For any nonnegative measurable function $f(x)$ in $(0, \infty)$, the following inequality holds true:

$$\left[\int_0^\infty y^{-p\sigma_1-1} \left(\int_0^\infty h\left(\frac{x}{y}\right) f(x) dx \right)^p dy \right]^{\frac{1}{p}}$$

$$\leq M \left[\int_0^\infty x^{p(1-\sigma)-1} f^p(x) dx \right]^{\frac{1}{p}}. \tag{2.23}$$

(ii) For any nonnegative measurable functions $f(x)$ and $G(y)$ in $(0, \infty)$, we have the following inequality:

$$\int_0^\infty \int_0^\infty h\left(\frac{x}{y}\right) f(x)G(y)dxdy$$
$$\leq M \left[\int_0^\infty x^{p(1-\sigma)-1} f^p(x)dx\right]^{\frac{1}{p}} \left[\int_0^\infty y^{q(1+\sigma_1)-1} G^q(y)dy\right]^{\frac{1}{q}}. \qquad (2.24)$$

(iii) For $k(\sigma) > 0$, we have $\sigma_1 = \sigma$, and $k(\sigma) \leq M \ (< \infty)$.

For $\sigma_1 = \sigma$, we have

Corollary 2.7 *Suppose that M is a constant. The following Statements (i), (ii) and (iii) are equivalent:*
(i) For any nonnegative measurable function $f(x)$ in $(0, \infty)$, we have the following inequality:

$$\left[\int_0^\infty y^{-p\sigma-1} \left(\int_0^\infty h\left(\frac{x}{y}\right) f(x)dx\right)^p dy\right]^{\frac{1}{p}}$$
$$\leq M \left[\int_0^\infty x^{p(1-\sigma)-1} f^p(x)dx\right]^{\frac{1}{p}}. \qquad (2.25)$$

(ii) For any nonnegative measurable functions $f(x)$ and $G(y)$ in $(0, \infty)$, we have the following inequality:

$$\int_0^\infty \int_0^\infty h\left(\frac{x}{y}\right) f(x)G(y)dxdy$$
$$\leq M \left[\int_0^\infty x^{p(1-\sigma)-1} f^p(x)dx\right]^{\frac{1}{p}} \left[\int_0^\infty y^{q(1+\sigma)-1} G^q(y)dy\right]^{\frac{1}{q}}. \qquad (2.26)$$

(iii) $k(\sigma) \leq M < \infty$.
Moreover, if (iii) holds true, then the constant factor $M = k(\sigma)$ in (2.25) and (2.26) is the best possible.

For $M = k(\sigma)$, we derive the following corollary:

Corollary 2.8 *Suppose that $k(\sigma) \in \mathbf{R}_+$. The following Statements (i) and (ii) are valid and equivalent:*
(i) For any $f(x) \geq 0$, satisfying

$$0 < \int_0^\infty x^{p(1-\sigma)-1} f^p(x)dx < \infty,$$

we have the following inequality:

$$\left[\int_0^\infty y^{-p\sigma-1} \left(\int_0^\infty h\left(\frac{x}{y} \right) f(x)dx \right)^p dy \right]^{\frac{1}{p}}$$

$$< k(\sigma) \left[\int_0^\infty x^{p(1-\sigma)-1} f^p(x)dx \right]^{\frac{1}{p}}. \tag{2.27}$$

(ii) For any $f(x) \geq 0$, satisfying

$$0 < \int_0^\infty x^{p(1-\sigma)-1} f^p(x)dx < \infty,$$

and $g(y) \geq 0$, satisfying

$$0 < \int_0^\infty y^{q(1+\sigma)-1} G^q(y)dy < \infty,$$

we have the following inequality:

$$\int_0^\infty \int_0^\infty h\left(\frac{x}{y} \right) f(x)g(y)dxdy$$

$$< k(\sigma) \left[\int_0^\infty x^{p(1-\sigma)-1} f^p(x)dx \right]^{\frac{1}{p}} \left[\int_0^\infty y^{q(1+\sigma)-1} G^q(y)dy \right]^{\frac{1}{q}}. \tag{2.28}$$

Moreover, the constant factor $k(\sigma)$ in (2.27) and (2.28) is the best possible.

Note. $h(\frac{x}{y})$ is a homogeneous function of degree 0, namely,

$$h\left(\frac{x}{y} \right) = k_0(x, y).$$

Setting
$$h(u) = k_\lambda(u, 1),$$

where $k_\lambda(x, y)$ is the homogeneous function of degree $-\lambda \in \mathbf{R}$, then for

$$g(y) = y^\lambda G(y) \text{ and } \mu_1 = \lambda - \sigma_1$$

in Corollaries 2.6, 2.7 and 2.8, we have:

Corollary 2.9 *Suppose that M is a constant. The following Statements (i), (ii) and (iii) are equivalent:*
    (i) For any nonnegative measurable function $f(x)$ in $(0, \infty)$, we have the following inequality:

$$\left[\int_0^\infty y^{p\mu_1 - 1} \left(\int_0^\infty k_\lambda(x, y) f(x) dx \right)^p dy \right]^{\frac{1}{p}}$$

$$\leq M \left[\int_0^\infty x^{p(1-\sigma)-1} f^p(x) dx \right]^{\frac{1}{p}}. \tag{2.29}$$

(ii) For any nonnegative measurable functions $f(x)$ and $g(y)$ in $(0, \infty)$, we have the following inequality:

$$\int_0^\infty \int_0^\infty k_\lambda(x, y) f(x) g(y) dx dy$$

$$\leq M \left[\int_0^\infty x^{p(1-\sigma)-1} f^p(x) dx \right]^{\frac{1}{p}} \left[\int_0^\infty y^{q(1-\mu_1)-1} g^q(y) dy \right]^{\frac{1}{q}}. \tag{2.30}$$

(iii) For

$$k_\lambda(\sigma) := \int_0^\infty k_\lambda(u, 1) u^{\sigma-1} du > 0,$$

we have

$$\mu_1 = \mu \quad and \quad k_\lambda(\sigma) \leq M \ (< \infty).$$

For $\mu_1 = \mu$, we obtain the following corollary:

Corollary 2.10 *Suppose that M is a constant. The following Statements (i), (ii) and (iii) are equivalent:*

(i) For any nonnegative measurable function $f(x)$ in $(0, \infty)$, we have the following inequality:

$$\left[\int_0^\infty y^{p\mu - 1} \left(\int_0^\infty k_\lambda(x, y) f(x) dx \right)^p dy \right]^{\frac{1}{p}}$$

$$\leq M \left[\int_0^\infty x^{p(1-\sigma)-1} f^p(x) dx \right]^{\frac{1}{p}}. \tag{2.31}$$

(ii) For any nonnegative measurable functions $f(x)$ and $g(y)$ in $(0, \infty)$, we have the following inequality:

$$\int_0^\infty \int_0^\infty k_\lambda(x, y) f(x) g(y) dx dy$$

$$\leq M \left[\int_0^\infty x^{p(1-\sigma)-1} f^p(x) dx \right]^{\frac{1}{p}} \left[\int_0^\infty y^{q(1-\mu)-1} g^q(y) dy \right]^{\frac{1}{q}}. \tag{2.32}$$

(iii) $k_\lambda(\sigma) \leq M \ (< \infty)$.

Moreover, if (iii) holds true, then the constant factor $M = k_\lambda(\sigma)$ in (2.31) and (2.32) is the best possible.

For $M = k_\lambda(\sigma)$, we have

Corollary 2.11 *Suppose that $k_\lambda(\sigma) \in \mathbf{R}_+$. The following Statements (i) and (ii) are valid and equivalent:*

(i) For any $f(x) \geq 0$, satisfying

$$0 < \int_0^\infty x^{p(1-\sigma)-1} f^p(x)dx < \infty,$$

we have the following inequality:

$$\left[\int_0^\infty y^{p\mu-1} \left(\int_0^\infty k_\lambda(x, y)f(x)dx \right)^p dy \right]^{\frac{1}{p}}$$

$$< k_\lambda(\sigma) \left[\int_0^\infty x^{p(1-\sigma)-1} f^p(x)dx \right]^{\frac{1}{p}}. \tag{2.33}$$

(ii) For any $f(x) \geq 0$, satisfying

$$0 < \int_0^\infty x^{p(1-\sigma)-1} f^p(x)dx < \infty,$$

and $g(y) \geq 0$, satisfying

$$0 < \int_0^\infty y^{q(1-\mu)-1} g^q(y)dy < \infty,$$

we have the following inequality:

$$\int_0^\infty \int_0^\infty k_\lambda(x, y)f(x)g(y)dxdy$$

$$< k_\lambda(\sigma) \left[\int_0^\infty x^{p(1-\sigma)-1} f^p(x)dx \right]^{\frac{1}{p}} \left[\int_0^\infty y^{q(1-\mu)-1} g^q(y)dy \right]^{\frac{1}{q}}. \tag{2.34}$$

Moreover, the constant factor $k_\lambda(\sigma)$ in (2.33) and (2.34) is the best possible.

In particular,
(i) for $\lambda = 1, \sigma = \frac{1}{q}, \mu = \frac{1}{p}$, we have the following equivalent inequalities with the best possible constant factor

$$k_1\left(\frac{1}{q}\right) = \int_0^\infty k_1(u, 1)u^{-\frac{1}{p}}du :$$

$$\left[\int_0^\infty \left(\int_0^\infty k_1(x, y)f(x)dx\right)^p dy\right]^{\frac{1}{p}} < k_1\left(\frac{1}{q}\right)\left(\int_0^\infty f^p(x)dx\right)^{\frac{1}{p}}, \quad (2.35)$$

$$\int_0^\infty \int_0^\infty k_1(x, y)f(x)g(y)dxdy$$

$$< k_1\left(\frac{1}{q}\right)\left(\int_0^\infty f^p(x)dx\right)^{\frac{1}{p}}\left(\int_0^\infty g^q(y)dy\right)^{\frac{1}{q}}; \quad (2.36)$$

(ii) for $\lambda = 1, \sigma = \frac{1}{p}, \mu = \frac{1}{q}$, we have the following equivalent inequalities with the best possible constant factor

$$k_1\left(\frac{1}{p}\right) = \int_0^\infty k_1(u, 1)u^{-\frac{1}{q}}du :$$

$$\left[\int_0^\infty y^{p-2}\left(\int_0^\infty k_1(x, y)f(x)dx\right)^p dy\right]^{\frac{1}{p}}$$

$$< k_1\left(\frac{1}{p}\right)\left(\int_0^\infty x^{p-2}f^p(x)dx\right)^{\frac{1}{p}}, \quad (2.37)$$

$$\int_0^\infty \int_0^\infty k_1(x, y)f(x)g(y)dxdy$$

$$< k_1\left(\frac{1}{p}\right)\left(\int_0^\infty x^{p-2}f^p(x)dx\right)^{\frac{1}{p}}\left(\int_0^\infty y^{q-2}g^q(y)dy\right)^{\frac{1}{q}}. \quad (2.38)$$

Note. If

$$\lambda = 0, \mu = -\sigma, k_0(x, y) = h\left(\frac{x}{y}\right), k_0(\sigma) = k(\sigma), g(y) = G(y),$$

then Corollary 2.10 reduces to Corollary 2.7.

2.3 Operator Expressions and a Few Examples

We set the following functions:

$$\varphi(x) := x^{p(1-\sigma)-1}, \quad \psi(y) := y^{q(1-\sigma)-1}, \quad \phi(y) := y^{q(1-\mu)-1},$$

wherefrom,
$$\psi^{1-p}(y) = y^{p\sigma-1}, \; \phi^{1-p}(y) = y^{p\mu-1} \; (x, y \in \mathbf{R}_+).$$

Define the following real normed linear spaces:

$$L_{p,\varphi}(\mathbf{R}_+) := \left\{ f : ||f||_{p,\varphi} := \left(\int_0^\infty \varphi(x)|f(x)|^p dx \right)^{\frac{1}{p}} < \infty \right\},$$

wherefrom,

$$L_{q,\psi}(\mathbf{R}_+) = \left\{ g : ||g||_{q,\psi} := \left(\int_0^\infty \psi(y)|g(y)|^q dy \right)^{\frac{1}{q}} < \infty \right\},$$

$$L_{q,\phi}(\mathbf{R}_+) = \left\{ g : ||g||_{q,\phi} := \left(\int_0^\infty \phi(y)|g(y)|^q dy \right)^{\frac{1}{q}} < \infty \right\},$$

$$L_{p,\psi^{1-p}}(\mathbf{R}_+) = \left\{ h : ||h||_{p,\psi^{1-p}} = \left(\int_0^\infty \psi^{1-p}(y)|h(y)|^p dy \right)^{\frac{1}{p}} < \infty \right\},$$

$$L_{q,\phi^{1-p}}(\mathbf{R}_+) = \left\{ h : ||h||_{p,\phi^{1-p}} = \left(\int_0^\infty \phi^{1-p}(y)|h(y)|^p dy \right)^{\frac{1}{p}} < \infty \right\}.$$

(a) In view of Theorem 2.4, for $f \in L_{p,\varphi}(\mathbf{R}_+)$, setting

$$h_1(y) := \int_0^\infty h(xy)f(x)dx \; (y \in \mathbf{R}_+),$$

by (2.17), we have

$$||h_1||_{p,\psi^{1-p}} = \left[\int_0^\infty \psi^{1-p}(y)h_1^p(y)dy \right]^{\frac{1}{p}} \le M||f||_{p,\varphi} < \infty. \tag{2.39}$$

Definition 2.12 Define a Hilbert-type integral operator with the nonhomogeneous kernel

$$T^{(1)} : L_{p,\varphi}(\mathbf{R}_+) \to L_{p,\psi^{1-p}}(\mathbf{R}_+)$$

as follows:
    For any $f \in L_{p,\varphi}(\mathbf{R}_+)$, there exists a unique representation

$$T^{(1)}f = h_1 \in L_{p,\psi^{1-p}}(\mathbf{R}_+),$$

satisfying

$$T^{(1)} f(y) = h_1(y),$$

for any $y \in \mathbf{R}_+$,

In view of (2.39), it follows that

$$\|T^{(1)} f\|_{p,\psi^{1-p}} = \|h_1\|_{p,\psi^{1-p}} \leq M \|f\|_{p,\varphi},$$

and then the operator $T^{(1)}$ is bounded satisfying

$$\|T^{(1)}\| = \sup_{f(\neq \theta) \in L_{p,\varphi}(\mathbf{R}_+)} \frac{\|T^{(1)} f\|_{p,\psi^{1-p}}}{\|f\|_{p,\varphi}} \leq M.$$

If we define the formal inner product of $T^{(1)} f$ and g as follows:

$$(T^{(1)} f, g) := \int_0^\infty \left(\int_0^\infty h(xy) f(x) dx \right) g(y) dy,$$

then we can rewrite Theorem 2.3 as follows:

Theorem 2.13 *Suppose that M is a constant. The following Statements (i), (ii) and (iii) are equivalent:*
(i) For any $f(x) \geq 0$, $f \in L_{p,\varphi}(\mathbf{R}_+)$, we have the following inequality:

$$\|T^{(1)} f\|_{p,\psi^{1-p}} \leq M \|f\|_{p,\varphi}. \tag{2.40}$$

(ii) For any $f(x), g(y) \geq 0$, $f \in L_{p,\varphi}(\mathbf{R}_+)$, $g \in L_{q,\psi}(\mathbf{R}_+)$, we have the following inequality:

$$(T^{(1)} f, g) \leq M \|f\|_{p,\varphi} \|g\|_{q,\psi}. \tag{2.41}$$

(iii) $k(\sigma) \leq M \ (< \infty)$.
Moreover, if Statement (iii) holds true, then the constant factor $M = k(\sigma)$ in (2.40) and (2.41) is the best possible, namely,

$$\|T^{(1)}\| = k(\sigma) \leq M.$$

(b) In view of Corollary 2.10, for $f \in L_{p,\varphi}(\mathbf{R}_+)$, setting

$$h_2(y) := \int_0^\infty k_\lambda(x, y) f(x) dx \ (y \in \mathbf{R}_+),$$

by (2.31), we have

$$\|h_2\|_{p,\phi^{1-p}} = \left[\int_0^\infty \phi^{1-p}(y) h_2^p(y) dy \right]^{\frac{1}{p}} < M \|f\|_{p,\varphi} < \infty. \tag{2.42}$$

Definition 2.14 Define a Hilbert-type integral operator with the homogeneous kernel

$$T^{(2)} : L_{p,\varphi}(\mathbf{R}_+) \to L_{p,\phi^{1-p}}(\mathbf{R}_+)$$

as follows:

For any $f \in L_{p,\varphi}(\mathbf{R})$, there exists a unique representation

$$T^{(2)}f = h_2 \in L_{p,\phi^{1-p}}(\mathbf{R}_+),$$

satisfying

$$T^{(2)}f(y) = h_2(y),$$

for any $y \in \mathbf{R}_+$.

In view of (2.42), it follows that

$$\|T^{(2)}f\|_{p,\phi^{1-p}} = \|h_2\|_{p,\phi^{1-p}} \le M\|f\|_{p,\varphi},$$

and then the operator $T^{(2)}$ is bounded satisfying

$$\|T^{(2)}\| = \sup_{f(\neq\theta)\in L_{p,\varphi}(\mathbf{R}_+)} \frac{\|T^{(2)}f\|_{p,\phi^{1-p}}}{\|f\|_{p,\varphi}} \le M.$$

If we define the formal inner product of $T^{(2)}f$ and g as follows:

$$(T^{(2)}f, g) := \int_0^\infty \left(\int_0^\infty k_\lambda(x, y) f(x)dx \right) g(y)dy,$$

then we can rewrite Corollary 2.10 in the following manner:

Corollary 2.15 *Suppose that M is a constant. The following statements (i), (ii) and (iii) are equivalent:*
   (i) For any $f(x) \ge 0$, $f \in L_{p,\varphi}(\mathbf{R}_+)$, we have the following inequality:

$$\|T^{(2)}f\|_{p,\phi^{1-p}} \le M\|f\|_{p,\varphi}. \tag{2.43}$$

   (ii) For any $f(x), g(y) \ge 0$, $f \in L_{p,\varphi}(\mathbf{R}_+)$, $g \in L_{q,\phi}(\mathbf{R}_+)$, we have the following inequality:

$$(T^{(2)}f, g) \le M\|f\|_{p,\varphi}\|g\|_{q,\phi}. \tag{2.44}$$

   (iii) $k_\lambda(\sigma) \le M$ ($< \infty$).
   Moreover, if statement (iii) holds true, then the constant factor $M = k_\lambda(\sigma)$ in (2.43) and (2.44) is the best possible, namely,

$$\|T^{(2)}\| = k_\lambda(\sigma) \le M.$$

Example 2.16 Setting

$$h(u) = k_0(u, 1) = csch(u) = \frac{2}{e^u - e^{-u}} \quad (u > 0),$$

where, $csch(u)$ is the hyperbolic cosecant function (cf. [3]), then we get

$$h(xy) = csch(xy) = \frac{2}{e^{xy} - e^{-xy}},$$

$$k_0(x, y) = csch\left(\frac{x}{y}\right) = \frac{2}{e^{x/y} - e^{-x/y}},$$

and for $\sigma > 1$, it follows that

$$k(\sigma) = k_0(\sigma) = \int_0^\infty csch(u)u^{\sigma-1}du$$

$$= \int_0^\infty \frac{2u^{\sigma-1}}{e^u - e^{-u}}du = \int_0^\infty \frac{2u^{\sigma-1}e^{-u}}{1 - e^{-2u}}du$$

$$= 2\int_0^\infty u^{\sigma-1}\sum_{k=0}^\infty e^{-(2k+1)u}du$$

$$= 2\sum_{k=0}^\infty \int_0^\infty u^{\sigma-1}e^{-(2k+1)u}du.$$

Setting $v = (2k + 1)u$ in the previous integral, we have

$$k(\sigma) = k_0(\sigma) = 2\int_0^\infty v^{\sigma-1}e^{-v}dv \sum_{k=0}^\infty \frac{1}{(2k + 1)^\sigma}$$

$$= 2\Gamma(\sigma)\left[\sum_{k=1}^\infty \frac{1}{k^\sigma} - \sum_{k=1}^\infty \frac{1}{(2k)^\sigma}\right]$$

$$= 2\Gamma(\sigma)\left(1 - \frac{1}{2^\sigma}\right)\varsigma(\sigma) \in \mathbf{R}_+,$$

where

$$\Gamma(s) := \int_0^\infty v^{s-1}e^{-s}dv \, (Res > 0)$$

is the gamma function, and

$$\varsigma(s) := \sum_{k=1}^\infty \frac{1}{k^s} \, (Res > 1)$$

is the Riemann zeta function (cf. [4]). Then by Theorem 2.13 and Corollary 2.15, we have

$$||T^{(1)}|| = ||T^{(2)}|| = 2\Gamma(\sigma)\left(1 - \frac{1}{2^{\sigma}}\right)\zeta(\sigma). \tag{2.45}$$

Example 2.17 Setting

$$h(u) = k_0(u, 1) = e^{-u}\,sech(u) = \frac{2e^{-u}}{e^u + e^{-u}}\ (u > 0),$$

where, $sech(u)$ is the hyperbolic secant function (cf. [3]), then we get

$$h(xy) = e^{-xy}\,sech(xy) = \frac{2e^{-xy}}{e^{xy} + e^{-xy}},$$

$$k_0(x, y) = e^{-x/y}\,sech(\frac{x}{y}) = \frac{2e^{-x/y}}{e^{x/y} + e^{-x/y}},$$

and for $\sigma > 1$, it follows that

$$
\begin{aligned}
k(\sigma) = k_0(\sigma) &= \int_0^\infty e^{-u}\,sech(u)u^{\sigma-1}du \\
&= \int_0^\infty \frac{2e^{-u}u^{\sigma-1}}{e^u + e^{-u}}du = 2\int_0^\infty \frac{u^{\sigma-1}e^{-2u}}{1 + e^{-2u}}du \\
&= 2\int_0^\infty u^{\sigma-1}\sum_{k=0}^\infty (-1)^k e^{-2(k+1)u}du \\
&= 2\sum_{k=0}^\infty (-1)^k \int_0^\infty u^{\sigma-1}e^{-2(k+1)u}du.
\end{aligned}
$$

Setting $v = 2(k + 1)u$ in the previous integral, we obtain that

$$
\begin{aligned}
k(\sigma) = k_0(\sigma) &= \frac{1}{2^{\sigma-1}}\int_0^\infty v^{\sigma-1}e^{-v}dv \sum_{k=1}^\infty \frac{(-1)^k}{k^\sigma} \\
&= \frac{1}{2^{\sigma-1}}\Gamma(\sigma)\left[\sum_{k=1}^\infty \frac{1}{k^\sigma} - 2\sum_{k=1}^\infty \frac{1}{(2k)^\sigma}\right] \\
&= \frac{1}{2^{\sigma-1}}\Gamma(\sigma)\left(1 - \frac{1}{2^{\sigma-1}}\right)\zeta(\sigma) \in \mathbf{R}_+.
\end{aligned}
$$

Then by Theorem 2.13 and Corollary 2.15, we have

$$||T^{(1)}|| = ||T^{(2)}|| = \frac{1}{2^{\sigma-1}}\Gamma(\sigma)\left(1 - \frac{1}{2^{\sigma-1}}\right)\zeta(\sigma). \tag{2.46}$$

Example 2.18 Setting

$$h(u) = k_\lambda(u, 1) = \frac{|\ln u|^\beta (\min\{u, 1\})^\alpha}{(\max\{u, 1\})^{\lambda+\alpha}} \quad (u > 0),$$

we then obtain that

$$h(xy) = \frac{|\ln xy|^\beta (\min\{xy, 1\})^\alpha}{(\max\{xy, 1\})^{\lambda+\alpha}},$$

$$k_\lambda(x, y) = \frac{|\ln x/y|^\beta (\min\{x, y\})^\alpha}{(\max\{x, y\})^{\lambda+\alpha}} \quad (x, y > 0)$$

and for $\beta > -1$, σ, $\mu > -\alpha$, it follows that

$$k(\sigma) = k_\lambda(\sigma) = \int_0^\infty \frac{|\ln u|^\beta (\min\{u, 1\})^\alpha}{(\max\{u, 1\})^{\lambda+\alpha}} u^{\sigma-1} du$$

$$= \int_0^1 (-\ln u)^\beta (u^{\alpha+\sigma-1} + u^{\alpha+\mu-1}) du$$

$$= \int_0^\infty v^\beta (e^{-(\alpha+\sigma)v} + e^{-(\alpha+\mu)v}) dv$$

$$= \frac{\Gamma(\beta+1)(\lambda+2\alpha)}{(\sigma+\alpha)(\mu+\alpha)} \in \mathbf{R}_+.$$

Then by Theorem 2.13 and Corollary 2.15, we have

$$||T^{(1)}|| = ||T^{(2)}|| = \frac{\Gamma(\beta+1)(\lambda+2\alpha)}{(\sigma+\alpha)(\mu+\alpha)}. \tag{2.47}$$

Example 2.19 Setting

$$h(u) = k_\lambda(u, 1) = \frac{|\ln u|^\beta}{|u^\lambda - 1|} \quad (u > 0),$$

we then derive that

$$h(xy) = \frac{|\ln xy|^\beta}{|(xy)^\lambda - 1|},$$

$$k_\lambda(x, y) = \frac{|\ln x/y|^\beta}{|x^\lambda - y^\lambda|} \quad (x, y > 0),$$

and for β, σ, $\mu > 0$, it follows that

$$k(\sigma) = k_\lambda(\sigma) = \int_0^\infty \frac{|\ln u|^\beta}{|u^\lambda - 1|} u^{\sigma-1} du$$

$$= \int_0^1 \frac{(-\ln u)^\beta}{1 - u^\lambda} u^{\sigma-1} du + \int_1^\infty \frac{\ln^\beta u}{u^\lambda - 1} u^{\sigma-1} du$$

$$= \int_0^1 \frac{(-\ln u)^\beta}{1 - u^\lambda} (u^{\sigma-1} + u^{\mu-1}) du$$

$$= \int_0^1 (-\ln u)^\beta \sum_{k=0}^\infty u^{\lambda k} (u^{\sigma-1} + u^{\mu-1}) du.$$

By the Lebesgue term by term integration theorem (cf. [1]), we have

$$k(\sigma) = k_\lambda(\sigma) = \sum_{k=0}^\infty \int_0^1 (-\ln u)^\beta (u^{\lambda k + \sigma - 1} + u^{\lambda k + \mu - 1}) du$$

$$= \sum_{k=0}^\infty \left[\frac{1}{(\lambda k + \sigma)^{\beta+1}} + \frac{1}{(\lambda k + \mu)^{\beta+1}} \right] \int_0^\infty v^\beta e^{-v} dv$$

$$= \frac{\Gamma(\beta + 1)}{\lambda^{\beta+1}} \left(\zeta(\beta + 1, \frac{\sigma}{\lambda}) + \zeta(\beta + 1, \frac{\mu}{\lambda}) \right) \in \mathbf{R}_+,$$

where

$$\zeta(s, a) = \sum_{k=0}^\infty \frac{1}{(k + a)^s} \quad (Res > 1; 0 < a \leq 1)$$

is the Hurwitz zeta function ($\zeta(s, 1) = \zeta(s)$ is the Riemann zeta function) (cf. [4]). Then by Theorem 2.13 and Corollary 2.15, we have

$$||T^{(1)}|| = ||T^{(2)}|| = \frac{\Gamma(\beta + 1)}{\lambda^{\beta+1}} \left(\zeta(\beta + 1, \frac{\sigma}{\lambda}) + \zeta(\beta + 1, \frac{\mu}{\lambda}) \right). \tag{2.48}$$

2.4　Introducing the Exponent Function as an Interval Variable

For $a, b \in \mathbf{R}\backslash\{0\}$, replacing x (resp. y) by e^{ax} (resp. e^{by}), then replacing $f(e^{ax})e^{ax}$ (resp. $g(e^{by})e^{by}$) by $f(x)$ (resp. $g(y)$), and $\frac{M}{|a|^{1/q}|b|^{1/p}}$ by M in Theorem 2.3, and by carrying out the corresponding simplifications, we derive the following theorem:

Theorem 2.20 *Suppose that M is a constant. The following statements (i), (ii) and (iii) are equivalent:*

(i) For any nonnegative measurable function $f(x)$ in $(-\infty, \infty)$, we have the following inequality:

$$\left[\int_{-\infty}^{\infty} e^{p\sigma_1 by} \left(\int_{-\infty}^{\infty} h(e^{ax+by}) f(x) dx\right)^p dy\right]^{\frac{1}{p}}$$

$$\leq M \left[\int_{-\infty}^{\infty} \left(\frac{f(x)}{e^{\sigma ax}}\right)^p dx\right]^{\frac{1}{p}}. \tag{2.49}$$

(ii) For any nonnegative measurable functions $f(x)$ and $g(y)$ in $(-\infty, \infty)$, we have the following inequality:

$$\int_{-\infty}^{\infty} \int_{-\infty}^{\infty} h(e^{ax+by}) f(x) g(y) dx dy$$

$$\leq M \left[\int_{-\infty}^{\infty} \left(\frac{f(x)}{e^{\sigma ax}}\right)^p dx\right]^{\frac{1}{p}} \left[\int_{-\infty}^{\infty} \left(\frac{g(y)}{e^{\sigma_1 by}}\right)^q dy\right]^{\frac{1}{q}}. \tag{2.50}$$

(iii) For $k(\sigma) > 0$, we have

$$\sigma_1 = \sigma \quad \text{and} \quad \frac{k(\sigma)}{|a|^{1/q}|b|^{1/p}} \leq M \quad (< \infty).$$

For $\sigma_1 = \sigma$ in *Theorem* 2.20, we have

Theorem 2.21 *Let M be a constant. The following statements (i), (ii) and (iii) are equivalent:*
(i) For any nonnegative measurable function $f(x)$ in $(-\infty, \infty)$, we have the following inequality:

$$\left[\int_{-\infty}^{\infty} e^{p\sigma by} \left(\int_{-\infty}^{\infty} h(e^{ax+by}) f(x) dx\right)^p dy\right]^{\frac{1}{p}}$$

$$\leq M \left[\int_{-\infty}^{\infty} \left(\frac{f(x)}{e^{\sigma ax}}\right)^p dx\right]^{\frac{1}{p}}. \tag{2.51}$$

(ii) For any nonnegative measurable functions $f(x)$ and $g(y)$ in $(-\infty, \infty)$, we have the following inequality:

$$\int_{-\infty}^{\infty} \int_{-\infty}^{\infty} h(e^{ax+by}) f(x) g(y) dx dy$$

$$\leq M \left[\int_{-\infty}^{\infty} \left(\frac{f(x)}{e^{\sigma ax}}\right)^p dx\right]^{\frac{1}{p}} \left[\int_{-\infty}^{\infty} \left(\frac{g(y)}{e^{\sigma by}}\right)^q dy\right]^{\frac{1}{q}}. \tag{2.52}$$

(iii)

$$\frac{k(\sigma)}{|a|^{1/q}|b|^{1/p}} \leq M \quad (< \infty).$$

Moreover, if statement (iii) holds true, then the constant factor $M = \frac{k(\sigma)}{|a|^{1/q}|b|^{1/p}}$ in (2.51) and (2.52) is the best possible.

For $M = \frac{k(\sigma)}{|a|^{1/q}|b|^{1/p}}$, we have

Theorem 2.22 *Suppose that $k(\sigma) \in \mathbf{R}_+$. The following statements (i) and (ii) are valid and equivalent:*

(i) For any $f(x) \geq 0$, satisfying

$$0 < \int_{-\infty}^{\infty} \left(\frac{f(x)}{e^{\sigma a x}}\right)^p dx < \infty,$$

we have the following inequality:

$$\left[\int_{-\infty}^{\infty} e^{p\sigma b y}\left(\int_{-\infty}^{\infty} h(e^{ax+by})f(x)dx\right)^p dy\right]^{\frac{1}{p}}$$

$$< \frac{k(\sigma)}{|a|^{1/q}|b|^{1/p}}\left[\int_{-\infty}^{\infty}\left(\frac{f(x)}{e^{\sigma a x}}\right)^p dx\right]^{\frac{1}{p}}. \tag{2.53}$$

(ii) For any $f(x) \geq 0$, satisfying

$$0 < \int_{-\infty}^{\infty}\left(\frac{f(x)}{e^{\sigma a x}}\right)^p dx < \infty,$$

and $g(y) \geq 0$, satisfying

$$0 < \int_{-\infty}^{\infty}\left(\frac{g(y)}{e^{\sigma b y}}\right)^q dy < \infty,$$

we have the following inequality:

$$\int_{-\infty}^{\infty}\int_{-\infty}^{\infty} h(e^{ax+by})f(x)g(y)dxdy$$

$$< \frac{k(\sigma)}{|a|^{1/q}|b|^{1/p}}\left[\int_{-\infty}^{\infty}\left(\frac{f(x)}{e^{\sigma a x}}\right)^p dx\right]^{\frac{1}{p}}\left[\int_{-\infty}^{\infty}\left(\frac{g(y)}{e^{\sigma b y}}\right)^q dy\right]^{\frac{1}{q}}. \tag{2.54}$$

Moreover, the constant factor

$$\frac{k(\sigma)}{|a|^{1/q}|b|^{1/p}}$$

in (2.53) and (2.54) is the best possible.

In particular, for

$$h(u) = e^{-u} \operatorname{sech}(u)(u > 0),$$

by Example 2.17, we have the following equivalent inequalities with the best possible constant factor

$$\frac{1}{2^{\sigma-1}}\Gamma(\sigma)\left(1-\frac{1}{2^{\sigma-1}}\right)\zeta(\sigma) \ (\sigma > 1):$$

$$\left[\int_{-\infty}^{\infty} e^{p\sigma by}\left(\int_{-\infty}^{\infty}\frac{sech(e^{ax+by})}{e^{e^{ax+by}}}f(x)dx\right)^p dy\right]^{\frac{1}{p}}$$

$$< \frac{2^{1-\sigma}\Gamma(\sigma)}{|a|^{1/q}|b|^{1/p}}\left(1-\frac{1}{2^{\sigma-1}}\right)\zeta(\sigma)\left[\int_{-\infty}^{\infty}\left(\frac{f(x)}{e^{\sigma ax}}\right)^p dx\right]^{\frac{1}{p}}, \qquad (2.55)$$

$$\int_{-\infty}^{\infty}\int_{-\infty}^{\infty}\frac{sech(e^{ax+by})}{e^{e^{ax+by}}}f(x)g(y)dxdy$$

$$< \frac{2^{1-\sigma}\Gamma(\sigma)}{|a|^{1/q}|b|^{1/p}}\left(1-\frac{1}{2^{\sigma-1}}\right)\zeta(\sigma)$$

$$\times\left[\int_{-\infty}^{\infty}\left(\frac{f(x)}{e^{\sigma ax}}\right)^p dx\right]^{\frac{1}{p}}\left[\int_{-\infty}^{\infty}\left(\frac{g(y)}{e^{\sigma by}}\right)^q dy\right]^{\frac{1}{q}}. \qquad (2.56)$$

For $a, b \in \mathbf{R}\backslash\{0\}$, replacing x (resp. y) by e^{ax} (resp. e^{by}), then replacing $f(e^{ax})e^{ax}$ (resp. $g(e^{by})e^{by}$) by $f(x)$ (resp. $g(y)$), and $\frac{M}{|a|^{1/q}|b|^{1/p}}$ by M in Corollary 2.9, and by carrying out the corresponding simplifications, we have:

Corollary 2.23 *Let M be a constant. The following statements (i), (ii) and (iii) are equivalent:*

(i) For any nonnegative measurable function $f(x)$ in $(-\infty, \infty)$, we have the following inequality:

$$\left[\int_{-\infty}^{\infty} e^{p\mu_1 by}\left(\int_{-\infty}^{\infty} k_\lambda(e^{ax}, e^{by})f(x)dx\right)^p dy\right]^{\frac{1}{p}}$$

$$\leq M\left[\int_{-\infty}^{\infty}\left(\frac{f(x)}{e^{\sigma ax}}\right)^p dx\right]^{\frac{1}{p}}. \qquad (2.57)$$

(ii) For any nonnegative measurable functions $f(x)$ and $g(y)$ in $(-\infty, \infty)$, we have the following inequality:

$$\int_{-\infty}^{\infty}\int_{-\infty}^{\infty} k_\lambda(e^{ax}, e^{by})f(x)g(y)dxdy$$

$$\leq M\left[\int_{-\infty}^{\infty}\left(\frac{f(x)}{e^{\sigma ax}}\right)^p dx\right]^{\frac{1}{p}}\left[\int_{-\infty}^{\infty}\left(\frac{g(y)}{e^{\mu_1 by}}\right)^q dy\right]^{\frac{1}{q}}. \qquad (2.58)$$

(iii) For

$$k_\lambda(\sigma) = \int_0^\infty k_\lambda(u, 1)u^{\sigma-1}du > 0,$$

we have

$$\mu_1 = \mu \ \ and \ \ \frac{k_\lambda(\sigma)}{|a|^{1/q}|b|^{1/p}} \le M(< \infty).$$

For $\mu_1 = \mu$, we have

Corollary 2.24 *Let M be a constant. The following statements (i), (ii) and (iii) are equivalent:*

(i) For any nonnegative measurable function $f(x)$ in $(-\infty, \infty)$, we have the following inequality:

$$\left[\int_{-\infty}^\infty e^{p\mu by} \left(\int_{-\infty}^\infty k_\lambda(e^{ax}, e^{by}) f(x)dx\right)^p dy\right]^{\frac{1}{p}}$$

$$\le M \left[\int_{-\infty}^\infty \left(\frac{f(x)}{e^{\sigma ax}}\right)^p dx\right]^{\frac{1}{p}}. \tag{2.59}$$

(ii) For any nonnegative measurable functions $f(x)$ and $g(y)$ in $(-\infty, \infty)$, we have the following inequality:

$$\int_{-\infty}^\infty \int_{-\infty}^\infty k_\lambda(e^{ax}, e^{by}) f(x)g(y)dxdy$$

$$\le M \left[\int_{-\infty}^\infty \left(\frac{f(x)}{e^{\sigma ax}}\right)^p dx\right]^{\frac{1}{p}} \left[\int_{-\infty}^\infty \left(\frac{g(y)}{e^{\mu by}}\right)^q dy\right]^{\frac{1}{q}}. \tag{2.60}$$

(iii)

$$\frac{k_\lambda(\sigma)}{|a|^{1/q}|b|^{1/p}} \le M < \infty.$$

Moreover, if statement (iii) holds true, then the constant factor $M = \frac{k_\lambda(\sigma)}{|a|^{1/q}|b|^{1/p}}$ in (2.59) and (2.60) is the best possible.

For $M = \frac{k_\lambda(\sigma)}{|a|^{1/q}|b|^{1/p}}$, we obtain the following corollary:

Corollary 2.25 *Suppose that $k_\lambda(\sigma) \in \mathbf{R}_+$. The following statements (i) and (ii) are valid and equivalent:*

(i) For any $f(x) \ge 0$, satisfying

$$0 < \int_{-\infty}^\infty \left(\frac{f(x)}{e^{\sigma ax}}\right)^p dx < \infty,$$

we have the following inequality:

$$\left[\int_{-\infty}^{\infty} e^{p\mu by} \left(\int_{-\infty}^{\infty} k_\lambda(e^{ax}, e^{by}) f(x) dx \right)^p dy \right]^{\frac{1}{p}}$$
$$< \frac{k_\lambda(\sigma)}{|a|^{1/q} |b|^{1/p}} \left[\int_{-\infty}^{\infty} \left(\frac{f(x)}{e^{\sigma ax}} \right)^p dx \right]^{\frac{1}{p}}. \tag{2.61}$$

(ii) For any $f(x) \geq 0$, satisfying

$$0 < \int_{-\infty}^{\infty} \left(\frac{f(x)}{e^{\sigma ax}} \right)^p dx < \infty,$$

and $g(y) \geq 0$, satisfying

$$0 < \int_{-\infty}^{\infty} \left(\frac{g(y)}{e^{\mu by}} \right)^q dy < \infty,$$

we have the following inequality:

$$\int_{-\infty}^{\infty} \int_{-\infty}^{\infty} k_\lambda(e^{ax}, e^{by}) f(x) g(y) dx dy$$
$$< \frac{k_\lambda(\sigma)}{|a|^{1/q} |b|^{1/p}} \left[\int_{-\infty}^{\infty} \left(\frac{f(x)}{e^{\sigma ax}} \right)^p dx \right]^{\frac{1}{p}} \left[\int_{-\infty}^{\infty} \left(\frac{g(y)}{e^{\mu by}} \right)^q dy \right]^{\frac{1}{q}}. \tag{2.62}$$

Moreover, the constant factor

$$\frac{k_\lambda(\sigma)}{|a|^{1/q} |b|^{1/p}}$$

in (2.61) and (2.62) is the best possible.

In particular, for

$$k_\lambda(x, y) = \frac{|\ln x/y|^\beta}{|x^\lambda - y^\lambda|},$$

by Example 2.19, we derive the following equivalent inequalities with the best possible constant factor

$$K_\lambda(\sigma) := \frac{\Gamma(\beta + 1)}{|a|^{1/q} |b|^{1/p} \lambda^{\beta+1}} \left(\zeta(\beta + 1, \frac{\sigma}{\lambda}) + \zeta(\beta + 1, \frac{\mu}{\lambda}) \right) \quad (\beta, \sigma, \mu > 0):$$

$$\left[\int_{-\infty}^{\infty} e^{p\mu by} \left(\int_{-\infty}^{\infty} \frac{|ax - by|^{\beta}}{|e^{\lambda ax} - e^{\lambda by}|} f(x) dx \right)^{p} dy \right]^{\frac{1}{p}}$$

$$< K_{\lambda}(\sigma) \left[\int_{-\infty}^{\infty} \left(\frac{f(x)}{e^{\sigma ax}} \right)^{p} dx \right]^{\frac{1}{p}}, \tag{2.63}$$

$$\int_{-\infty}^{\infty} \int_{-\infty}^{\infty} \frac{|ax - by|^{\beta}}{|e^{\lambda ax} - e^{\lambda by}|} f(x) g(y) dx dy$$

$$< K_{\lambda}(\sigma) \left[\int_{-\infty}^{\infty} \left(\frac{f(x)}{e^{\sigma ax}} \right)^{p} dx \right]^{\frac{1}{p}} \left[\int_{-\infty}^{\infty} \left(\frac{g(y)}{e^{\mu by}} \right)^{q} dy \right]^{\frac{1}{q}}. \tag{2.64}$$

References

1. Kuang, J.C.: Real and Functional Analysis (continuation) (sec. vol.). Higher Education Press, Beijing, China (2015)
2. Kuang, J.C.: Applied Inequalities. Shangdong Science and Technology Press, Jinan, China (2004)
3. Zhong, Y.Q.: Introduction to Complex Functions (Third Volume). Higher Education Press, Beijing (2003)
4. Wang, Z.X., Guo, D.R.: Introduction to Special Functions. Science Press, Beijing (1979)

Chapter 3
Equivalent Statements of the Reverse Hilbert-Type Integral Inequalities

In this chapter, by the use of weight functions, a few statements of the reverse Hilbert-type integral inequalities with a general nonhomogeneous kernel related to certain parameters are obtained. Additionally, some equivalent statements of the reverse Hilbert-type integral inequalities with the general homogeneous kernel are deduced. We also consider some particular cases, a few examples related to the extended Hurwitz zeta function in the form of applications, as well as the case of the reverse Hilbert-type integral inequalities in the whole plane.

3.1 Some Lemmas

Throughout this chapter we shall assume that: $0 < p < 1$ $(q < 0)$, $\frac{1}{p} + \frac{1}{q} = 1$, $\sigma_1, \mu, \sigma \in \mathbf{R}$, $\mu + \sigma = \lambda$, and $h(u)$ is a nonnegative measurable function in $(0, \infty)$, such that

$$k(\sigma) = \int_0^\infty h(u)u^{\sigma-1}du (\geq 0) . \tag{3.1}$$

For $n \in \mathbf{N}$, we consider the following two expressions:

$$I_1 = \int_1^\infty \left(\int_0^1 h(xy)x^{\sigma+\frac{1}{pn}-1}dx \right) y^{\sigma_1-\frac{1}{qn}-1}dy, \tag{3.2}$$

$$I_2 = \int_0^1 \left(\int_1^\infty h(xy)x^{\sigma-\frac{1}{pn}-1}dx \right) y^{\sigma_1+\frac{1}{qn}-1}dy. \tag{3.3}$$

© The Author(s), under exclusive licence to Springer Nature Switzerland AG 2019
B. Yang and M. Th. Rassias, *On Hilbert-Type and Hardy-Type Integral Inequalities and Applications*, SpringerBriefs in Mathematics,
https://doi.org/10.1007/978-3-030-29268-3_3

Setting $u = xy$ in (3.2) and (3.3), we obtain

$$I_1 = \int_1^\infty \left[\int_0^y h(u) \left(\frac{u}{y} \right)^{\sigma + \frac{1}{pn} - 1} \frac{1}{y} du \right] y^{\sigma_1 - \frac{1}{qn} - 1} dy$$

$$= \int_1^\infty y^{(\sigma_1 - \sigma) - \frac{1}{n} - 1} \left(\int_0^y h(u) u^{\sigma + \frac{1}{pn} - 1} du \right) dy, \tag{3.4}$$

$$I_2 = \int_0^1 \left[\int_y^\infty h(u) \left(\frac{u}{y} \right)^{\sigma - \frac{1}{pn} - 1} \frac{1}{y} du \right] y^{\sigma_1 + \frac{1}{qn} - 1} dy$$

$$= \int_0^1 y^{(\sigma_1 - \sigma) + \frac{1}{n} - 1} \left(\int_y^\infty h(u) u^{\sigma - \frac{1}{pn} - 1} du \right) dy. \tag{3.5}$$

Lemma 3.1 *If $k(\sigma) < \infty$ and there exist constants $\delta_0, M > 0$ such that $k(\sigma \pm \delta_0) < \infty$, and if for any nonnegative measurable functions $f(x)$ and $g(y)$ in $(0, \infty)$, the following inequality*

$$I = \int_0^\infty \int_0^\infty h(xy) f(x) g(y) dx dy$$

$$\geq M \left[\int_0^\infty x^{p(1-\sigma)-1} f^p(x) dx \right]^{\frac{1}{p}} \left[\int_0^\infty y^{q(1-\sigma_1)-1} g^q(y) dy \right]^{\frac{1}{q}} \tag{3.6}$$

holds true, then we have $\sigma_1 = \sigma$ and $k(\sigma) \geq M \, (> 0)$.

Proof If $\sigma_1 > \sigma$, then for $n > \frac{1}{\delta_0 p}$ $(n \in \mathbf{N})$, we consider the following two functions:

$$f_n(x) = \begin{cases} 0, & 0 < x < 1 \\ x^{\sigma - \frac{1}{pn} - 1}, & x \geq 1 \end{cases},$$

$$g_n(y) = \begin{cases} y^{\sigma_1 + \frac{1}{qn} - 1}, & 0 < y \leq 1 \\ 0, & y > 1 \end{cases}.$$

We obtain that

$$J_2 = \left[\int_0^\infty x^{p(1-\sigma)-1} f_n^p(x) dx \right]^{\frac{1}{p}} \left[\int_0^\infty y^{q(1-\sigma_1)-1} g_n^q(y) dy \right]^{\frac{1}{q}}$$

$$= \left(\int_1^\infty x^{-\frac{1}{n} - 1} dx \right)^{\frac{1}{p}} \left(\int_0^1 y^{\frac{1}{n} - 1} dy \right)^{\frac{1}{q}} = n.$$

By (3.5), we have

$$I_2 \le \int_0^1 y^{(\sigma_1-\sigma)+\frac{1}{n}-1} dy \int_0^\infty h(u)u^{\sigma-\frac{1}{pn}-1} du$$

$$= \frac{1}{\sigma_1 - \sigma + \frac{1}{n}}$$

$$\times \left(\int_0^1 h(u)u^{\sigma-\frac{1}{pn}-1} du + \int_1^\infty h(u)u^{\sigma-\frac{1}{pn}-1} du \right)$$

$$\le \frac{1}{\sigma_1 - \sigma} \left(\int_0^1 h(u)u^{(\sigma-\delta_0)-1} du + \int_1^\infty h(u)u^{\sigma-1} du \right)$$

$$\le \frac{1}{\sigma_1 - \sigma} \left(k(\sigma - \delta_0) + k(\sigma) \right),$$

and then by (3.6), it follows that

$$\frac{1}{\sigma_1 - \sigma} \left(k(\sigma - \delta_0) + k(\sigma) \right)$$

$$\ge I_2 = \int_0^\infty \int_0^\infty h(xy) f_n(x) g_n(y) dx dy \ge M J_2 = Mn. \tag{3.7}$$

By (3.7), in view of $\sigma_1 - \sigma > 0$, and

$$0 \le k(\sigma - \delta_0) + k(\sigma) < \infty,$$

for $n \to \infty$, we find that

$$\infty > \frac{1}{\sigma_1 - \sigma} \left(k(\sigma - \delta_0) + k(\sigma) \right) \ge \infty,$$

which is a contradiction.

If $\sigma_1 < \sigma$, then for $n \in \mathbf{N}, n > \frac{1}{\delta_0 p}$, we consider the following two functions:

$$\tilde{f}_n(x) = \begin{cases} x^{\sigma+\frac{1}{pn}-1}, & 0 < x \le 1 \\ 0, & x > 1 \end{cases},$$

$$\tilde{g}_n(y) = \begin{cases} 0, & 0 < y < 1 \\ y^{\sigma_1-\frac{1}{qn}-1}, & y \ge 1 \end{cases}.$$

We get that

$$\tilde{J}_2 = \left[\int_0^\infty x^{p(1-\sigma)-1} \tilde{f}_n^p(x) dx \right]^{\frac{1}{p}} \left[\int_0^\infty y^{q(1-\sigma_1)-1} \tilde{g}_n^q(y) dy \right]^{\frac{1}{q}}$$

$$= \left(\int_0^1 x^{\frac{1}{n}-1} dx \right)^{\frac{1}{p}} \left(\int_1^\infty y^{-\frac{1}{n}-1} dy \right)^{\frac{1}{q}} = n.$$

By (3.4), we have

$$I_1 \leq \int_1^\infty y^{(\sigma_1-\sigma)-\frac{1}{n}-1} dy \int_0^\infty h(u) u^{\sigma+\frac{1}{pn}-1} du$$

$$= \frac{1}{\sigma - \sigma_1 + \frac{1}{n}}$$

$$\times \left(\int_0^1 h(u) u^{\sigma+\frac{1}{pn}-1} du + \int_1^\infty h(u) u^{\sigma+\frac{1}{pn}-1} du \right)$$

$$\leq \frac{1}{\sigma - \sigma_1} \left(\int_0^1 h(u) u^{\sigma-1} du + \int_1^\infty h(u) u^{\sigma+\delta_0-1} du \right)$$

$$\leq \frac{1}{\sigma - \sigma_1} (k(\sigma) + k(\sigma + \delta_0)),$$

and then by (3.6), it follows that

$$\frac{1}{\sigma - \sigma_1} (k(\sigma) + k(\sigma + \delta_0))$$

$$\geq I_1 = \int_0^\infty \int_0^\infty h(xy) \widetilde{f_n}(x) \widetilde{g_n}(y) dx dy \geq M \widetilde{J_2} = Mn. \tag{3.8}$$

By (2.8), for $n \to \infty$, we also obtain that

$$\infty > \frac{1}{\sigma - \sigma_1} (k(\sigma) + k(\sigma + \delta_0)) \geq \infty,$$

which is a contradiction. Hence, we conclude that $\sigma_1 = \sigma$.

For $\sigma_1 = \sigma$, we have
$$nM = M J_2 \leq I_2$$

by the use of (3.6), and we reduce (3.5) as follows:

$$M = \frac{1}{n} M J_2 \leq \frac{1}{n} I_2$$

$$= \frac{1}{n} \int_0^1 y^{\frac{1}{n}-1} \left(\int_y^\infty h(u) u^{\sigma-\frac{1}{pn}-1} du \right) dy$$

$$= \frac{1}{n} \int_0^1 y^{\frac{1}{n}-1} \left(\int_y^1 h(u) u^{\sigma-\frac{1}{pn}-1} du \right) dy + \int_1^\infty h(u) u^{\sigma-\frac{1}{pn}-1} du$$

$$= \frac{1}{n} \int_0^1 \left(\int_0^u y^{\frac{1}{n}-1} dy \right) h(u) u^{\sigma-\frac{1}{pn}-1} du + \int_1^\infty h(u) u^{\sigma-\frac{1}{pn}-1} du$$

$$\leq \int_0^1 h(u) u^{\sigma+\frac{1}{qn}-1} du + \int_1^\infty h(u) u^{\sigma-1} du. \tag{3.9}$$

Since for $n > \frac{1}{\delta_0 |q|}$ $(n \in \mathbf{N})$ we have

$$h(u)u^{\sigma + \frac{1}{qn} - 1} \le h(u)u^{\sigma - \delta_0 - 1} \quad (0 < u \le 1)$$

and

$$\int_0^1 h(u)u^{\sigma - \delta_0 - 1} du \le k(\sigma - \delta_0) < \infty,$$

by (3.9) and the Lebesgue dominated convergence theorem (cf. [2]), we obtain

$$k(\sigma) = \int_0^1 \lim_{n \to \infty} h(u)u^{\sigma + \frac{1}{qn} - 1} du + \int_1^\infty h(u)u^{\sigma - 1} du$$

$$= \lim_{n \to \infty} \left[\int_0^1 h(u)u^{\sigma + \frac{1}{qn} - 1} du + \int_1^\infty h(u)u^{\sigma - 1} du \right] \ge M.$$

This completes the proof of the lemma. □

For $\sigma_1 = \sigma$, we have:

Lemma 3.2 *If $k(\sigma) < \infty$ and there exist constants δ_0, $M > 0$ such that $k(\sigma - \delta_0) < \infty$ or $k(\sigma + \delta_0) < \infty$, and if for any nonnegative measurable functions $f(x)$ and $g(y)$ in $(0, \infty)$ the following inequality*

$$I := \int_0^\infty \int_0^\infty h(xy) f(x) g(y) dx dy$$

$$\ge M \left[\int_0^\infty x^{p(1-\sigma)-1} f^p(x) dx \right]^{\frac{1}{p}} \left[\int_0^\infty y^{q(1-\sigma)-1} g^q(y) dy \right]^{\frac{1}{q}} \quad (3.10)$$

holds true, then we have $k(\sigma) \ge M \ (> 0)$.

Proof If $k(\sigma - \delta_0) < \infty$, then for $\sigma_1 = \sigma$, by (3.9) and the proof of Lemma 3.1, we have

$$k(\sigma) \ge M.$$

If $k(\sigma + \delta_0) < \infty$, then we have

$$nM = M \widetilde{J}_2 \le I_1 \quad (\text{for } \sigma_1 = \sigma),$$

by the use of (3.10), and we reduce (3.4) as follows:

$$M = \frac{1}{n}M\tilde{J}_2 \leq \frac{1}{n}I_1$$

$$= \frac{1}{n}\int_1^\infty y^{-\frac{1}{n}-1}\left(\int_0^y h(u)u^{\sigma+\frac{1}{pn}-1}du\right)dy$$

$$= \int_0^1 h(u)u^{\sigma+\frac{1}{pn}-1}du + \frac{1}{n}\int_1^\infty y^{-\frac{1}{n}-1}\left(\int_1^y h(u)u^{\sigma+\frac{1}{pn}-1}du\right)dy$$

$$= \int_0^1 h(u)u^{\sigma+\frac{1}{pn}-1}du + \frac{1}{n}\int_1^\infty \left(\int_u^\infty y^{-\frac{1}{n}-1}dy\right)h(u)u^{\sigma+\frac{1}{pn}-1}du$$

$$\leq \int_0^1 h(u)u^{\sigma-1}du + \int_1^\infty h(u)u^{\sigma-\frac{1}{qn}-1}du. \tag{3.11}$$

Since for $n > \frac{1}{\delta_0|q|}$ ($n \in \mathbf{N}$) we have

$$h(u)u^{\sigma-\frac{1}{qn}-1} \leq h(u)u^{\sigma+\delta_0-1} \ (u \geq 1)$$

and

$$\int_1^\infty h(u)u^{\sigma+\delta_0-1}du \leq k(\sigma+\delta_0) < \infty,$$

by (3.11) and the Lebesgue dominated convergence theorem (cf. [2]), we have

$$k(\sigma) = \int_0^1 h(u)u^{\sigma-1}du + \int_1^\infty \lim_{n\to\infty} h(u)u^{\sigma-\frac{1}{qn}-1}du$$

$$= \lim_{n\to\infty}\left[\int_0^1 h(u)u^{\sigma-1}du + \int_1^\infty h(u)u^{\sigma-\frac{1}{qn}-1}du\right] \geq M(> 0).$$

This completes the proof of the lemma.   □

3.2 Main Results

Theorem 3.3 *Suppose that M is a positive constant. The following statements (i),
(ii), (iii) and (iv) are equivalent:*
   (i) For any $f(x) \geq 0$, satisfying

$$0 < \int_0^\infty x^{p(1-\sigma)-1}f^p(x)dx < \infty,$$

we have the following inequality:

$$J := \left[\int_0^\infty y^{p\sigma_1 - 1} \left(\int_0^\infty h(xy)f(x)dx \right)^p dy \right]^{\frac{1}{p}}$$

$$> M \left[\int_0^\infty x^{p(1-\sigma)-1} f^p(x)dx \right]^{\frac{1}{p}}. \tag{3.12}$$

(ii) For any $g(y) \geq 0$, satisfying

$$0 < \int_0^\infty y^{q(1-\sigma_1)-1} g^q(y)dy < \infty,$$

we have the following inequality:

$$L := \left[\int_0^\infty x^{q\sigma - 1} \left(\int_0^\infty h(xy)g(y)dy \right)^q dx \right]^{\frac{1}{q}}$$

$$> M \left[\int_0^\infty y^{q(1-\sigma_1)-1} g^q(y)dy \right]^{\frac{1}{q}}. \tag{3.13}$$

(iii) For any $f(x) \geq 0$, satisfying

$$0 < \int_0^\infty x^{p(1-\sigma)-1} f^p(x)dx < \infty,$$

and $g(y) \geq 0$, satisfying

$$0 < \int_0^\infty y^{q(1-\sigma_1)-1} g^q(y)dy < \infty,$$

we have the following inequality:

$$I = \int_0^\infty \int_0^\infty h(xy)f(x)g(y)dxdy$$

$$> M \left[\int_0^\infty x^{p(1-\sigma)-1} f^p(x)dx \right]^{\frac{1}{p}} \left[\int_0^\infty y^{q(1-\sigma_1)-1} g^q(y)dy \right]^{\frac{1}{q}}. \tag{3.14}$$

(iv) For $k(\sigma) < \infty$, if there exists a constant $\delta_0 > 0$ such that $k(\sigma \pm \delta_0) < \infty$, then we have $\sigma_1 = \sigma$, and $k(\sigma) \geq M \ (> 0)$.

Proof (i) $\Rightarrow$ (iii). By the reverse Hölder inequality (cf. [1]), we have

$$I = \int_0^\infty \left(y^{\sigma_1 - \frac{1}{p}} \int_0^\infty h(xy)f(x)dx \right) \left(y^{\frac{1}{p} - \sigma_1} g(y) \right) dy$$

$$\geq J \left[\int_0^\infty y^{q(1-\sigma_1)-1} g^q(y)dy \right]^{\frac{1}{q}}. \tag{3.15}$$

Then by (3.12), we deduce (3.14).

$(ii) \Rightarrow (iii)$. By the reverse Hölder inequality, we also get that

$$I = \int_0^\infty \left(x^{\frac{1}{q}-\sigma} f(x)\right) \left(x^{\sigma-\frac{1}{q}} \int_0^\infty h(xy)g(y)dy\right) dx$$

$$\geq \left[\int_0^\infty x^{p(1-\sigma)-1} f^p(x)dx\right]^{\frac{1}{p}} L. \tag{3.16}$$

Then by (3.13), we deduce (3.14).

$(iii) \Rightarrow (iv)$. By Lemma 3.1, we have $\sigma_1 = \sigma$, and $k(\sigma) \geq M \ (> 0)$.

$(iv) \Rightarrow (i)$. For $y > 0$, setting $u = xy$, we obtain the following weight function:

$$\omega(\sigma, y) := y^\sigma \int_0^\infty h(xy)x^{\sigma-1}dx = \int_0^\infty h(u)u^{\sigma-1}du = k(\sigma) \ (y \in \mathbf{R}_+). \tag{3.17}$$

By the reverse Hölder inequality with weight and (3.17), we obtain that

$$\left(\int_0^\infty h(xy)f(x)dx\right)^p$$

$$= \left\{\int_0^\infty h(xy)\left[\frac{y^{(\sigma-1)/p}}{x^{(\sigma-1)/q}}f(x)\right]\left[\frac{x^{(\sigma-1)/q}}{y^{(\sigma-1)/p}}\right]dx\right\}^p$$

$$\geq \int_0^\infty h(xy)\frac{y^{\sigma-1}}{x^{(\sigma-1)p/q}}f^p(x)dx \left[\int_0^\infty h(xy)\frac{x^{\sigma-1}}{y^{(\sigma-1)q/p}}dx\right]^{p/q}$$

$$= \left[\omega(\sigma, y)y^{q(1-\sigma)-1}\right]^{p-1}\int_0^\infty h(xy)\frac{y^{\sigma-1}}{x^{(\sigma-1)p/q}}f^p(x)dx$$

$$= (k(\sigma))^{p-1}y^{-p\sigma+1}\int_0^\infty h(xy)\frac{y^{\sigma-1}}{x^{(\sigma-1)p/q}}f^p(x)dx. \tag{3.18}$$

If (3.18) assumes the form of equality for some $y \in (0, \infty)$, then (cf. [1]) there exist constants A and B, such that they are not all zero, and

$$A\frac{y^{\sigma-1}}{x^{(\sigma-1)p/q}}f^p(x) = B\frac{x^{\sigma-1}}{y^{(\sigma-1)q/p}} \quad a.e. \text{ in } \mathbf{R}_+.$$

Let us assume that $A \neq 0$ (otherwise $B = A = 0$). Then it follows that

$$x^{p(1-\sigma)-1}f^p(x) = y^{q(1-\sigma)}\frac{B}{Ax} \quad a.e. \text{ in } \mathbf{R}_+,$$

which contradicts the fact that

$$0 < \int_0^\infty x^{p(1-\sigma)-1}f^p(x)dx < \infty.$$

Hence, (3.18) takes the form of strict inequality.

For $\sigma_1 = \sigma$, by (3.18) and Fubini's theorem, we have

$$J > (k(\sigma))^{\frac{1}{q}} \left[\int_0^\infty \int_0^\infty h(xy) \frac{y^{\sigma-1}}{x^{(\sigma-1)p/q}} f^p(x) dx dy \right]^{\frac{1}{p}}$$

$$= (k(\sigma))^{\frac{1}{q}} \left\{ \int_0^\infty \left[\int_0^\infty h(xy) \frac{y^{\sigma-1}}{x^{(\sigma-1)(p-1)}} dy \right] f^p(x) dx \right\}^{\frac{1}{p}}$$

$$= (k(\sigma))^{\frac{1}{q}} \left[\int_0^\infty \omega(\sigma, x) x^{p(1-\sigma)-1} f^p(x) dx \right]^{\frac{1}{p}}$$

$$= k(\sigma) \left[\int_0^\infty x^{p(1-\sigma)-1} f^p(x) dx \right]^{\frac{1}{p}}. \tag{3.19}$$

For $k(\sigma) \geq M > 0$, (3.12) follows.

$(iii) \Rightarrow (ii)$. We set

$$f(x) := x^{q\sigma-1} \left(\int_0^\infty h(xy) g(y) dy \right)^{q-1}, \quad x > 0.$$

If $L = \infty$, then (3.13) is trivially valid; if $L = 0$, then it is impossible. Suppose that $0 < L < \infty$. By (3.14), we have

$$\infty > \int_0^\infty x^{p(1-\sigma)-1} f^p(x) dx = L^q = I$$

$$> M \left[\int_0^\infty x^{p(1-\sigma)-1} f^p(x) dx \right]^{\frac{1}{p}} \left[\int_0^\infty y^{q(1-\sigma_1)-1} g^q(y) dy \right]^{\frac{1}{q}} > 0,$$

$$L = \left[\int_0^\infty x^{p(1-\sigma)-1} f^p(x) dx \right]^{\frac{1}{q}} > M \left[\int_0^\infty y^{q(1-\sigma_1)-1} g^q(y) dy \right]^{\frac{1}{q}},$$

namely, (3.13) follows.

Therefore, the statements (i), (ii), (iii) and (iv) are equivalent.

This completes the proof of the theorem. □

For $\sigma_1 = \sigma$, we also have the following theorem:

Theorem 3.4 *If $0 < k(\sigma) < \infty$, then the following statements (i), (ii) and (iii) are equivalent:*

(i) For any $f(x) \geq 0$, satisfying

$$0 < \int_0^\infty x^{p(1-\sigma)-1} f^p(x) dx < \infty,$$

we have the following inequality:

$$\left[\int_0^\infty y^{p\sigma-1}\left(\int_0^\infty h(xy)f(x)dx\right)^p dy\right]^{\frac{1}{p}}$$

$$> k(\sigma)\left[\int_0^\infty x^{p(1-\sigma)-1}f^p(x)dx\right]^{\frac{1}{p}}. \tag{3.20}$$

(ii) For any $g(y) \geq 0$, satisfying

$$0 < \int_0^\infty y^{q(1-\sigma)-1}g^q(y)dy < \infty,$$

we have the following inequality:

$$\left[\int_0^\infty x^{q\sigma-1}\left(\int_0^\infty h(xy)g(y)dy\right)^q dx\right]^{\frac{1}{q}}$$

$$> k(\sigma)\left[\int_0^\infty y^{q(1-\sigma)-1}g^q(y)dy\right]^{\frac{1}{q}}. \tag{3.21}$$

(iii) For any $f(x) \geq 0$, satisfying

$$0 < \int_0^\infty x^{p(1-\sigma)-1}f^p(x)dx < \infty,$$

and $g(y) \geq 0$, satisfying

$$0 < \int_0^\infty y^{q(1-\sigma)-1}g^q(y)dy < \infty,$$

we have the following inequality:

$$\int_0^\infty \int_0^\infty h(xy)f(x)g(y)dxdy$$

$$> k(\sigma)\left[\int_0^\infty x^{p(1-\sigma)-1}f^p(x)dx\right]^{\frac{1}{p}}\left[\int_0^\infty y^{q(1-\sigma)-1}g^q(y)dy\right]^{\frac{1}{q}}. \tag{3.22}$$

Moreover, if there exists a constant $\delta_0 > 0$ such that

$$k(\sigma - \delta_0) < \infty \text{ or } k(\sigma + \delta_0) < \infty,$$

then the constant factor $k(\sigma)$ in (3.20), (3.21) and (3.22) is the best possible.

Proof For $\sigma_1 = \sigma$ in Theorem 3.3, since $0 < k(\sigma) < \infty$, setting $M = k(\sigma)$ in (3.12), (3.13) and (3.14), we can similarly prove that statements (i), (ii) and (iii) of Theorem 3.4 are equivalent.

If there exists a constant $M \geq k(\sigma)$, such that (3.22) is valid, then by Lemma 3.2, we have $M \leq k(\sigma)$. Hence, the constant factor $M = k(\sigma)$ in (3.22) is the best possible. The constant factor $k(\sigma)$ in (3.20) (3.21) is still the best possible. Otherwise, by (3.15) ((3.16)) for $\sigma_1 = \sigma$, we can conclude that the constant factor $M = k(\sigma)$ in (3.22) is not the best possible.

This completes the proof of the theorem. □

3.3 Some Corollaries and a Few Examples

In particular, for $\sigma = \frac{1}{p} (> 1)$ in Theorem 3.4, we have:

Corollary 3.5 *If*

$$0 < k\left(\frac{1}{p}\right) = \int_0^\infty h(u)u^{-\frac{1}{q}}du < \infty,$$

then the following statements (i), (ii) and (iii) are valid and equivalent:
  (i) For any $f(x) \geq 0$, satisfying

$$0 < \int_0^\infty x^{p-2} f^p(x)dx < \infty,$$

we have the following inequality:

$$\left[\int_0^\infty \left(\int_0^\infty h(xy)f(x)dx\right)^p dy\right]^{\frac{1}{p}} > k\left(\frac{1}{p}\right)\left(\int_0^\infty x^{p-2}f^p(x)dx\right)^{\frac{1}{p}}. \quad (3.23)$$

  (ii) For any $g(y) \geq 0$, satisfying

$$0 < \int_0^\infty g^q(y)dy < \infty,$$

we have the following inequality:

$$\left[\int_0^\infty x^{q-2}\left(\int_0^\infty h(xy)g(y)dy\right)^q dx\right]^{\frac{1}{q}} > k\left(\frac{1}{p}\right)\left(\int_0^\infty g^q(y)dy\right)^{\frac{1}{q}}. \quad (3.24)$$

  (iii) For any $f(x) \geq 0$, satisfying

$$0 < \int_0^\infty x^{p-2}f^p(x)dx < \infty,$$

and $g(y) \geq 0$, satisfying

$$0 < \int_0^\infty g^q(y)dy < \infty,$$

we have the following inequality:

$$\int_0^\infty \int_0^\infty h(xy) f(x) g(y) dx dy$$

$$> k\left(\frac{1}{p}\right) \left(\int_0^\infty x^{p-2} f^p(x) dx\right)^{\frac{1}{p}} \left(\int_0^\infty g^q(y) dy\right)^{\frac{1}{q}}. \tag{3.25}$$

Moreover, if there exists a constant $\delta_0 > 0$ such that

$$k\left(\frac{1}{p} - \delta_0\right) < \infty \ \text{ or } \ k\left(\frac{1}{p} + \delta_0\right) < \infty,$$

then the constant factor $k(\frac{1}{p})$ in (3.23), (3.24) and (3.25) is the best possible.

In particular, for $\sigma = \frac{1}{q} \ (< 0)$ in Theorem 3.4, we derive the following corollary:

Corollary 3.6 *If*

$$0 < k\left(\frac{1}{q}\right) = \int_0^\infty h(u) u^{-\frac{1}{p}} du < \infty,$$

then the following statements (i), (ii) and (iii) are equivalent:
(i) For any $f(x) \geq 0$, satisfying

$$0 < \int_0^\infty f^p(x) dx < \infty,$$

we have the following inequality:

$$\left[\int_0^\infty y^{p-2} \left(\int_0^\infty h(xy) f(x) dx\right)^p dy\right]^{\frac{1}{p}} > k\left(\frac{1}{q}\right) \left(\int_0^\infty f^p(x) dx\right)^{\frac{1}{p}}. \tag{3.26}$$

(ii) For any $g(y) \geq 0$, satisfying

$$0 < \int_0^\infty y^{q-2} g^q(y) dy < \infty,$$

we have the following inequality:

$$\left[\int_0^\infty \left(\int_0^\infty h(xy) g(y) dy\right)^q dx\right]^{\frac{1}{q}} > k\left(\frac{1}{q}\right) \left(\int_0^\infty y^{q-2} g^q(y) dy\right)^{\frac{1}{q}}. \tag{3.27}$$

(iii) For any $f(x) \geq 0$, satisfying

$$0 < \int_0^\infty f^p(x) dx < \infty,$$

and $g(y) \geq 0$, satisfying

$$0 < \int_0^\infty y^{q-2} g^q(y) dy < \infty,$$

we have the following inequality:

$$\int_0^\infty \int_0^\infty h(xy) f(x) g(y) dx dy$$

$$> k \left(\frac{1}{q} \right) \left(\int_0^\infty f^p(x) dx \right)^{\frac{1}{p}} \left(\int_0^\infty y^{q-2} g^q(y) dy \right)^{\frac{1}{q}}. \qquad (3.28)$$

Moreover, if there exists a constant $\delta_0 > 0$, such that

$$k \left(\frac{1}{q} - \delta_0 \right) < \infty \ \text{or} \ k \left(\frac{1}{q} + \delta_0 \right) < \infty,$$

then the constant factor $k(\frac{1}{q})$ in (3.26), (3.27) and (3.28) is the best possible.

Setting

$$y = \frac{1}{Y}, \quad G(Y) = g \left(\frac{1}{Y} \right) \frac{1}{Y^2}$$

in Theorem 3.3, then replacing Y by y, we obtain the following corollary:

Corollary 3.7 *Let M be a positive constant. The following statements (i), (ii) and (iii) are equivalent:*
(i) For any $f(x) \geq 0$, satisfying

$$0 < \int_0^\infty x^{p(1-\sigma)-1} f^p(x) dx < \infty,$$

we have the following inequality:

$$\left[\int_0^\infty y^{-p\sigma_1-1} \left(\int_0^\infty h \left(\frac{x}{y} \right) f(x) dx \right)^p dy \right]^{\frac{1}{p}}$$

$$> M \left[\int_0^\infty x^{p(1-\sigma)-1} f^p(x) dx \right]^{\frac{1}{p}}. \qquad (3.29)$$

(ii) For any $G(y) \geq 0$, satisfying

$$0 < \int_0^\infty y^{q(1+\sigma_1)-1} G^q(y) dy < \infty,$$

we have the following inequality:

$$\left[\int_0^\infty x^{q\sigma-1}\left(\int_0^\infty h\left(\frac{x}{y}\right)G(y)dy\right)^q dx\right]^{\frac{1}{q}}$$

$$> M\left[\int_0^\infty y^{q(1+\sigma_1)-1}G^q(y)dy\right]^{\frac{1}{q}}. \tag{3.30}$$

(iii) For any $f(x) \geq 0$, satisfying

$$0 < \int_0^\infty x^{p(1-\sigma)-1}f^p(x)dx < \infty,$$

and $G(y) \geq 0$, satisfying

$$0 < \int_0^\infty y^{q(1+\sigma_1)-1}G^q(y)dy < \infty,$$

we have the following inequality:

$$\int_0^\infty \int_0^\infty h\left(\frac{x}{y}\right)f(x)G(y)dxdy$$

$$< M\left[\int_0^\infty x^{p(1-\sigma)-1}f^p(x)dx\right]^{\frac{1}{p}}\left[\int_0^\infty y^{q(1+\sigma_1)-1}G^q(y)dy\right]^{\frac{1}{q}}. \tag{3.31}$$

(iv) For $k(\sigma) < \infty$, if there exists a constant $\delta_0 > 0$ such that $k(\sigma \pm \delta_0) < \infty$, then $\sigma_1 = \sigma$, and $k(\sigma) \geq M\ (> 0)$.

Note. $h(\frac{x}{y})$ is a homogeneous function of degree 0, namely, we may write

$$h\left(\frac{x}{y}\right) = k_0(x, y).$$

Setting $h(u) = k_\lambda(u, 1)$, where $k_\lambda(x, y)$ is the homogeneous function of degree $-\lambda \in \mathbf{R}$, then for

$$g(y) = y^\lambda G(y) \quad \text{and} \quad \mu_1 = \lambda - \sigma_1$$

in Corollary 3.7, we obtain the following:

Corollary 3.8 *Let M be a positive constant. If*

$$k_\lambda(\sigma) = \int_0^\infty k_\lambda(u, 1)u^{\sigma-1}du < \infty,$$

then the following statements (i), (ii), (iii) and (iv) are equivalent:
(i) For any $f(x) \geq 0$, satisfying

$$0 < \int_0^\infty x^{p(1-\sigma)-1}f^p(x)dx < \infty,$$

we have the following inequality:

$$\left[\int_0^\infty y^{p\mu_1-1}\left(\int_0^\infty k_\lambda(x,y)f(x)dx\right)^p dy\right]^{\frac{1}{p}}$$
$$> M\left[\int_0^\infty x^{p(1-\sigma)-1}f^p(x)dx\right]^{\frac{1}{p}};\qquad(3.32)$$

(ii) For any $g(y) \geq 0$, satisfying

$$0 < \int_0^\infty y^{q(1-\mu_1)-1}g^q(y)dy < \infty,$$

we have the following inequality:

$$\left[\int_0^\infty x^{q\sigma-1}\left(\int_0^\infty k_\lambda(x,y)g(y)dy\right)^q dx\right]^{\frac{1}{q}}$$
$$> M\left[\int_0^\infty y^{q(1-\mu_1)-1}g^q(y)dy\right]^{\frac{1}{q}}.\qquad(3.33)$$

(iii) For any $f(x) \geq 0$, satisfying

$$0 < \int_0^\infty x^{p(1-\sigma)-1}f^p(x)dx < \infty,$$

and $g(y) \geq 0$, satisfying

$$0 < \int_0^\infty y^{q(1-\mu_1)-1}g^q(y)dy < \infty,$$

we have the following inequality:

$$\int_0^\infty\int_0^\infty k_\lambda(x,y)f(x)g(y)dxdy$$
$$> M\left[\int_0^\infty x^{p(1-\sigma)-1}f^p(x)dx\right]^{\frac{1}{p}}\left[\int_0^\infty y^{q(1-\mu_1)-1}g^q(y)dy\right]^{\frac{1}{q}}.\qquad(3.34)$$

(iv) If there exists a constant $\delta_0 > 0$, such that $k_\lambda(\sigma \pm \delta_0) < \infty$, then

$$\mu_1 = \mu \quad and \quad k_\lambda(\sigma) \geq M \ (> 0).$$

For $\mu_1 = \mu$ in Corollary 3.8, we also derive the corollary bellow:

Corollary 3.9 *If*

$$0 < k_\lambda(\sigma) < \infty,$$

then the following statements (i), (ii) and (iii) are equivalent:
(i) For any $f(x) \geq 0$, satisfying

$$0 < \int_0^\infty x^{p(1-\sigma)-1} f^p(x)dx < \infty,$$

we have the following inequality:

$$\left[\int_0^\infty y^{p\mu-1}\left(\int_0^\infty k_\lambda(x,y)f(x)dx\right)^p dy\right]^{\frac{1}{p}}$$

$$> k_\lambda(\sigma)\left[\int_0^\infty x^{p(1-\sigma)-1} f^p(x)dx\right]^{\frac{1}{p}}. \tag{3.35}$$

(ii) For any $g(y) \geq 0$, satisfying

$$0 < \int_0^\infty y^{q(1-\mu)-1} g^q(y)dy < \infty,$$

we have the following inequality:

$$\left[\int_0^\infty x^{q\sigma-1}\left(\int_0^\infty k_\lambda(x,y)g(y)dy\right)^q dx\right]^{\frac{1}{q}}$$

$$> k_\lambda(\sigma)\left[\int_0^\infty y^{q(1-\mu)-1} g^q(y)dy\right]^{\frac{1}{q}}. \tag{3.36}$$

(iii) For any $f(x) \geq 0$, satisfying

$$0 < \int_0^\infty x^{p(1-\sigma)-1} f^p(x)dx < \infty,$$

and $g(y) \geq 0$, satisfying

$$0 < \int_0^\infty y^{q(1-\mu)-1} g^q(y)dy < \infty,$$

we have the following inequality:

$$\int_0^\infty \int_0^\infty k_\lambda(x,y)f(x)g(y)dxdy$$

$$> k_\lambda(\sigma)\left[\int_0^\infty x^{p(1-\sigma)-1} f^p(x)dx\right]^{\frac{1}{p}}\left[\int_0^\infty y^{q(1-\mu)-1} g^q(y)dy\right]^{\frac{1}{q}}. \tag{3.37}$$

Moreover, if there exists a constant $\delta_0 > 0$, such that

$$k_\lambda(\sigma - \delta_0) < \infty \ \ or \ \ k_\lambda(\sigma + \delta_0) < \infty,$$

then the constant factor $k_\lambda(\sigma)$ in (3.35), (3.36) and (3.37) is the best possible.

In particular, for $\lambda = 1$, $\sigma = \frac{1}{q}$ (< 0), $\mu = \frac{1}{p}$ in Corollary 3.9, we derive the following corollary:

Corollary 3.10 *If*

$$0 < k_1\left(\frac{1}{q}\right) = \int_0^\infty k_1(u, 1)u^{-\frac{1}{p}}du < \infty,$$

then the following statements (i), (ii) and (iii) are equivalent:
(i) For any $f(x) \geq 0$, satisfying

$$0 < \int_0^\infty f^p(x)dx < \infty,$$

we have the following inequality:

$$\left[\int_0^\infty \left(\int_0^\infty k_1(x, y)f(x)dx\right)^p dy\right]^{\frac{1}{p}} > k_1\left(\frac{1}{q}\right)\left(\int_0^\infty f^p(x)dx\right)^{\frac{1}{p}}. \quad (3.38)$$

(ii) For any $g(y) \geq 0$, satisfying

$$0 < \int_0^\infty g^q(y)dy < \infty,$$

we have the following inequality:

$$\left[\int_0^\infty \left(\int_0^\infty k_1(x, y)g(y)dy\right)^q dx\right]^{\frac{1}{q}} > k_1\left(\frac{1}{q}\right)\left(\int_0^\infty g^q(y)dy\right)^{\frac{1}{q}}. \quad (3.39)$$

(iii) For any $f(x) \geq 0$, satisfying

$$0 < \int_0^\infty f^p(x)dx < \infty,$$

and $g(y) \geq 0$, satisfying

$$0 < \int_0^\infty g^q(y)dy < \infty,$$

we have the following inequality:

$$\int_0^\infty \int_0^\infty k_1(x, y) f(x) g(y) dx dy$$

$$> k_1 \left(\frac{1}{q}\right) \left(\int_0^\infty f^p(x) dx\right)^{\frac{1}{p}} \left(\int_0^\infty g^q(y) dy\right)^{\frac{1}{q}}. \qquad (3.40)$$

Moreover, if there exists a constant $\delta_0 > 0$, such that

$$k_1 \left(\frac{1}{q} - \delta_0\right) < \infty \text{ or } k_1 \left(\frac{1}{q} + \delta_0\right) < \infty,$$

then the constant factor $k_1(\frac{1}{q})$ in (3.38), (3.39) and (3.40) is the best possible.

For $\lambda = 1$, $\sigma = \frac{1}{p} (> 1)$, $\mu = \frac{1}{q}$ in Corollary 3.10, we can also deduce the corollary bellow:

Corollary 3.11 *If*

$$0 < k_1 \left(\frac{1}{p}\right) = \int_0^\infty k_1(u, 1) u^{-\frac{1}{q}} du < \infty,$$

then the following statements (i), (ii) and (iii) are equivalent:
(i) For any $f(x) \geq 0$, satisfying

$$0 < \int_0^\infty x^{p-2} f^p(x) dx < \infty,$$

we have the following inequality:

$$\left[\int_0^\infty y^{p-2} \left(\int_0^\infty k_1(x, y) f(x) dx\right)^p dy\right]^{\frac{1}{p}}$$

$$> k_1 \left(\frac{1}{p}\right) \left(\int_0^\infty x^{p-2} f^p(x) dx\right)^{\frac{1}{p}}. \qquad (3.41)$$

(ii) For any $g(y) \geq 0$, satisfying

$$0 < \int_0^\infty y^{q-2} g^q(y) dy < \infty,$$

we have the following inequality:

$$\left[\int_0^\infty x^{q-2} \left(\int_0^\infty k_1(x, y) g(y) dy\right)^q dx\right]^{\frac{1}{q}}$$

$$> k_1 \left(\frac{1}{p}\right) \left(\int_0^\infty y^{q-2} g^q(y) dy\right)^{\frac{1}{q}}. \qquad (3.42)$$

(iii) For any $f(x) \geq 0$, satisfying

$$0 < \int_0^\infty x^{p-2} f^p(x)dx < \infty,$$

and $g(y) \geq 0$, satisfying

$$0 < \int_0^\infty y^{q-2} g^q(y)dy < \infty,$$

we have the following inequality:

$$\int_0^\infty \int_0^\infty k_1(x, y) f(x)g(y)dxdy$$

$$> k_1\left(\frac{1}{p}\right) \left(\int_0^\infty x^{p-2} f^p(x)dx\right)^{\frac{1}{p}} \left(\int_0^\infty y^{q-2} g^q(y)dy\right)^{\frac{1}{q}}. \qquad (3.43)$$

Moreover, if there exists a constant $\delta_0 > 0$, such that

$$k_1\left(\frac{1}{p} - \delta_0\right) < \infty, \quad or \quad k_1\left(\frac{1}{p} + \delta_0\right) < \infty,$$

then the constant factor $k_1(\frac{1}{p})$ in (3.41), (3.42) and (3.43) is the best possible.

Example 3.12 Setting

$$h(u) = k_0(u, 1) = e^{\alpha u} \csc h(u) = \frac{2e^{\alpha u}}{e^u - e^{-u}} \ (u > 0),$$

where $csch(u)$ stands for the hyperbolic cosecant function, then we get that

$$h(xy) = e^{\alpha xy} csch(xy) = \frac{2e^{\alpha xy}}{e^{xy} - e^{-xy}},$$

$$k_0(x, y) = e^{\frac{\alpha x}{y}} \csc h(\frac{x}{y}) = \frac{2e^{\alpha x/y}}{e^{x/y} - e^{-x/y}},$$

and for $\alpha < 1, \sigma > 1$,

$$k(\sigma) = k_0(\sigma) = \int_0^\infty e^{\alpha u} csch(u)u^{\sigma-1}du$$

$$= \int_0^\infty \frac{2e^{\alpha u}u^{\sigma-1}}{e^u - e^{-u}}du = \int_0^\infty \frac{2u^{\sigma-1}e^{(\alpha-1)u}}{1 - e^{-2u}}du$$

$$= 2\int_0^\infty u^{\sigma-1} \sum_{k=0}^\infty e^{-(2k+1-\alpha)u}du$$

$$= 2\sum_{k=0}^\infty \int_0^\infty u^{\sigma-1}e^{-(2k+1-\alpha)u}du.$$

Setting $v = (2k + 1 - \alpha)u$ in the previous integral, we obtain that

$$k(\sigma) = k_0(\sigma) = 2 \int_0^\infty v^{\sigma-1} e^{-v} dv \sum_{k=0}^\infty \frac{1}{(2k + 1 - \alpha)^\sigma}$$

$$= \frac{1}{2^{\sigma-1}} \Gamma(\sigma) \sum_{k=0}^\infty \frac{1}{(k + \frac{1-\alpha}{2})^\sigma}$$

$$= \frac{1}{2^{\sigma-1}} \Gamma(\sigma) \zeta(\sigma, \frac{1 - \alpha}{2}) \in \mathbf{R}_+.$$

Setting $\delta_0 = \frac{\sigma-1}{2} > 0$,

$$\sigma \pm \delta_0 \geq \sigma - \frac{\sigma - 1}{2} = \frac{\sigma + 1}{2} > 1,$$

we have

$$k(\sigma \pm \delta_0) = k_0(\sigma \pm \delta_0) < \infty,$$

and then we can use the fact that

$$h(u) = k_0(u, 1) = e^{\alpha u} csch(u)$$

in Theorem 3.4 and Corollary 3.9 in order to obtain some equivalent inequalities with the best possible constant factor.

Example 3.13 Setting

$$h(u) = k_0(u, 1) = \frac{1}{|u - 1|^\lambda} (u > 0),$$

then we find

$$h(xy) = \frac{1}{|xy - 1|^\lambda}, \quad k_\lambda(x, y) = \frac{1}{|x - y|^\lambda},$$

and for $\sigma, \mu > 0, 0 < \lambda < 1$,

$$k(\sigma) = k_\lambda(\sigma) = \int_0^\infty \frac{1}{|u - 1|^\lambda} u^{\sigma-1} du$$

$$= \int_0^1 \frac{u^{\sigma-1}}{(1 - u)^\lambda} du + \int_1^\infty \frac{u^{\sigma-1}}{(u - 1)^\lambda} du$$

$$= \int_0^1 \frac{u^{\sigma-1}}{(1 - u)^\lambda} du + \int_0^1 \frac{v^{\mu-1}}{(1 - v)^\lambda} dv$$

$$= B(1 - \lambda, \sigma) + B(1 - \lambda, \mu) \in \mathbf{R}_+.$$

Setting $\delta_0 = \min\{\frac{\sigma}{2}, \frac{\mu}{2}\} > 0$,

$$\sigma \pm \delta_0 \geq \sigma - \frac{\sigma}{2} = \frac{\sigma}{2} > 0,$$

$$\mu \pm \delta_0 \geq \mu - \frac{\mu}{2} = \frac{\mu}{2} > 0,$$

we have

$$k(\sigma \pm \delta_0) = k_\lambda(\sigma \pm \delta_0) < \infty,$$

and then we can use the fact that

$$h(u) = k_\lambda(u, 1) = \frac{1}{|u - 1|^\lambda}$$

in Theorem 3.4 and Corollary 3.9 in order to obtain some equivalent inequalities with the best possible constant factor.

Example 3.14 Setting

$$h(u) = k_\lambda(u, 1) = \frac{1}{(u + 1)^\lambda} \ (u > 0),$$

we then get that

$$h(xy) = \frac{1}{(xy + 1)^\lambda}, k_\lambda(x, y) = \frac{1}{(x + y)^\lambda},$$

and for $\sigma, \mu > 0$,

$$k(\sigma) = k_\lambda(\sigma) = \int_0^\infty \frac{1}{(u + 1)^\lambda} u^{\sigma - 1} du$$
$$= B(\sigma, \mu) \in \mathbf{R}_+.$$

Setting $\delta_0 = \min\{\frac{\sigma}{2}, \frac{\mu}{2}\} > 0$,

$$\sigma \pm \delta_0 \geq \sigma - \frac{\sigma}{2} = \frac{\sigma}{2} > 0,$$

$$\mu \pm \delta_0 \geq \mu - \frac{\mu}{2} = \frac{\mu}{2} > 0,$$

we have

$$k(\sigma \pm \delta_0) = k_\lambda(\sigma \pm \delta_0) < \infty,$$

and we can then use the fact that

$$h(u) = k_\lambda(u, 1) = \frac{1}{(u + 1)^\lambda}$$

in Theorem 3.4 and Corollary 3.9 in order to obtain some equivalent inequalities with the best possible constant factor.

3.4 Some Reverse Equivalent Hilbert-Type Inequalities in the Whole Plane

For $a, b \in \mathbf{R}\backslash\{0\}$, if we replace x (resp. y) by e^{ax} (resp. e^{by}), then replacing $f(e^{ax})e^{ax}$ (resp. $g(e^{by})e^{by}$) by $f(x)$ (resp. $g(y)$), and $\frac{M}{|a|^{1/q}|b|^{1/p}}$ by M in Theorem 3.3, and by carrying out the corresponding simplifications, we get the following theorem:

Theorem 3.15 *Let M be a positive constant. The following statements (i), (ii), (iii) and (iv) are equivalent:*

(i) For any $f(x) \geq 0$, satisfying

$$0 < \int_{-\infty}^{\infty} \left(\frac{f(x)}{e^{\sigma a x}}\right)^p dx < \infty,$$

we have the following inequality:

$$\left[\int_{-\infty}^{\infty} e^{p\sigma_1 by} \left(\int_{-\infty}^{\infty} h(e^{ax+by})f(x)dx\right)^p dy\right]^{\frac{1}{p}}$$

$$> M \left[\int_{-\infty}^{\infty} \left(\frac{f(x)}{e^{\sigma a x}}\right)^p dx\right]^{\frac{1}{p}}. \tag{3.44}$$

(ii) For any $g(y) \geq 0$, satisfying

$$0 < \int_{-\infty}^{\infty} \left(\frac{g(y)}{e^{\sigma_1 by}}\right)^q dy < \infty$$

we have the following inequality:

$$\left[\int_{-\infty}^{\infty} e^{q\sigma a x} \left(\int_{-\infty}^{\infty} h(e^{ax+by})g(y)dy\right)^q dx\right]^{\frac{1}{q}}$$

$$> M \left[\int_{-\infty}^{\infty} \left(\frac{g(y)}{e^{\sigma_1 by}}\right)^q dy\right]^{\frac{1}{p}}. \tag{3.45}$$

(iii) For any $f(x) \geq 0$, satisfying

$$0 < \int_{-\infty}^{\infty} \left(\frac{f(x)}{e^{\sigma a x}}\right)^p dx < \infty$$

and $g(y) \geq 0$, satisfying

$$0 < \int_{-\infty}^{\infty} \left(\frac{g(y)}{e^{\sigma_1 by}}\right)^q dy < \infty,$$

we have the following inequality:

$$\int_{-\infty}^{\infty} \int_{-\infty}^{\infty} h(e^{ax+by}) f(x) g(y) dx dy$$

$$> M \left[\int_{-\infty}^{\infty} \left(\frac{f(x)}{e^{\sigma ax}} \right)^p dx \right]^{\frac{1}{p}} \left[\int_{-\infty}^{\infty} \left(\frac{g(y)}{e^{\sigma_1 by}} \right)^q dy \right]^{\frac{1}{q}}. \tag{3.46}$$

(iv) For $k(\sigma) < \infty$, *if there exists a constant* $\delta_0 > 0$, *such that* $k(\sigma \pm \delta_0) < \infty$, *then we have*

$$\sigma_1 = \sigma \text{ and } \frac{k(\sigma)}{|a|^{1/q} |b|^{1/p}} \geq M \ (> 0).$$

For $\sigma_1 = \sigma$, we derive the following theorem:

Theorem 3.16 *If* $0 < k(\sigma) < \infty$, *then the following statements (i), (ii) and (iii) are equivalent:*
  (i) For any $f(x) \geq 0$, *satisfying*

$$0 < \int_{-\infty}^{\infty} \left(\frac{f(x)}{e^{\sigma ax}} \right)^p dx < \infty,$$

we have the following inequality:

$$\left[\int_{-\infty}^{\infty} e^{p\sigma by} \left(\int_{-\infty}^{\infty} h(e^{ax+by}) f(x) dx \right)^p dy \right]^{\frac{1}{p}}$$

$$> \frac{k(\sigma)}{|a|^{1/q} |b|^{1/p}} \left[\int_{-\infty}^{\infty} \left(\frac{f(x)}{e^{\sigma ax}} \right)^p dx \right]^{\frac{1}{p}}. \tag{3.47}$$

(ii) For any $g(y) \geq 0$, *satisfying*

$$0 < \int_{-\infty}^{\infty} \left(\frac{g(y)}{e^{\sigma by}} \right)^q dy < \infty,$$

we have the following inequality:

$$\left[\int_{-\infty}^{\infty} e^{q\sigma ax} \left(\int_{-\infty}^{\infty} h(e^{ax+by}) g(y) dy \right)^q dx \right]^{\frac{1}{q}}$$

$$> \frac{k(\sigma)}{|a|^{1/q} |b|^{1/p}} \left[\int_{-\infty}^{\infty} \left(\frac{g(y)}{e^{\sigma by}} \right)^q dy \right]^{\frac{1}{p}}. \tag{3.48}$$

(iii) For any $f(x) \geq 0$, *satisfying*

$$0 < \int_{-\infty}^{\infty} \left(\frac{f(x)}{e^{\sigma a x}} \right)^p dx < \infty,$$

and $g(y) \geq 0$, satisfying

$$0 < \int_{-\infty}^{\infty} \left(\frac{g(y)}{e^{\sigma b y}} \right)^q dy < \infty,$$

we have the following inequality:

$$\int_{-\infty}^{\infty} \int_{-\infty}^{\infty} h(e^{ax+by}) f(x) g(y) dx dy$$

$$> \frac{k(\sigma)}{|a|^{1/q} |b|^{1/p}} \left[\int_{-\infty}^{\infty} \left(\frac{f(x)}{e^{\sigma a x}} \right)^p dx \right]^{\frac{1}{p}} \left[\int_{-\infty}^{\infty} \left(\frac{g(y)}{e^{\sigma b y}} \right)^q dy \right]^{\frac{1}{q}}. \qquad (3.49)$$

Moreover, if there exists a constant $\delta_0 > 0$, such that

$$k(\sigma - \delta_0) < \infty \text{ or } k(\sigma + \delta_0) < \infty,$$

then the constant factor

$$\frac{k(\sigma)}{|a|^{1/q} |b|^{1/p}}$$

in (3.47), (3.48) and (3.49) is the best possible.

In particular, for

$$h(u) = e^{\alpha u} csch(u),$$

by Example 3.12, we have the following equivalent inequalities with the best possible constant factor

$$\frac{2^{1-\sigma} \Gamma(\sigma)}{|a|^{1/q} |b|^{1/p}} \zeta \left(\sigma, \frac{1-\alpha}{2} \right)$$

$(\alpha < 1, \sigma > 1)$:

$$\left[\int_{-\infty}^{\infty} e^{p\sigma b y} \left(\int_{-\infty}^{\infty} e^{\alpha(e^{ax+by})} csch(e^{ax+by}) f(x) dx \right)^p dy \right]^{\frac{1}{p}}$$

$$> \frac{2^{1-\sigma} \Gamma(\sigma)}{|a|^{1/q} |b|^{1/p}} \zeta \left(\sigma, \frac{1-\alpha}{2} \right) \left[\int_{-\infty}^{\infty} \left(\frac{f(x)}{e^{\sigma a x}} \right)^p dx \right]^{\frac{1}{p}}, \qquad (3.50)$$

$$\left[\int_{-\infty}^{\infty} e^{q\sigma ax}\left(\int_{-\infty}^{\infty} e^{\alpha(e^{ax+by})}csch(e^{ax+by})g(y)dy\right)^q dx\right]^{\frac{1}{q}}$$

$$> \frac{2^{1-\sigma}\Gamma(\sigma)}{|a|^{1/q}|b|^{1/p}}\zeta\left(\sigma, \frac{1-\alpha}{2}\right)\left[\int_{-\infty}^{\infty}\left(\frac{g(y)}{e^{\sigma by}}\right)^q dy\right]^{\frac{1}{p}}, \qquad (3.51)$$

$$\int_0^{\infty}\int_0^{\infty} e^{\alpha(e^{ax+by})}csch(e^{ax+by})f(x)g(y)dxdy$$

$$> \frac{2^{1-\sigma}\Gamma(\sigma)}{|a|^{1/q}|b|^{1/p}}\zeta\left(\sigma, \frac{1-\alpha}{2}\right)$$

$$\times\left[\int_{-\infty}^{\infty}\left(\frac{f(x)}{e^{\sigma ax}}\right)^p dx\right]^{\frac{1}{p}}\left[\int_{-\infty}^{\infty}\left(\frac{g(y)}{e^{\sigma by}}\right)^q dy\right]^{\frac{1}{q}}. \qquad (3.52)$$

For $a, b \in \mathbf{R}\backslash\{0\}$, replacing x (resp. y) by e^{ax} (resp. e^{by}), then replacing $f(e^{ax})e^{ax}$ (resp. $g(e^{by})e^{by}$) by $f(x)$ (resp. $g(y)$), and $\frac{M}{|a|^{1/q}|b|^{1/p}}$ by M in Corollary 3.8, and by carrying out the corresponding simplifications, we get the following corollary:

Corollary 3.17 *Suppose that M is a positive constant. The following statements (i), (ii), (iii) and (iv) are equivalent:*
(i) For any $f(x) \geq 0$, satisfying

$$0 < \int_{-\infty}^{\infty}\left(\frac{f(x)}{e^{\sigma ax}}\right)^p dx < \infty,$$

we have the following inequality:

$$\left[\int_{-\infty}^{\infty} e^{p\mu_1 by}\left(\int_{-\infty}^{\infty} k_\lambda(e^{ax}, e^{by})f(x)dx\right)^p dy\right]^{\frac{1}{p}}$$

$$> M\left[\int_{-\infty}^{\infty}\left(\frac{f(x)}{e^{\sigma ax}}\right)^p dx\right]^{\frac{1}{p}}. \qquad (3.53)$$

(ii) For any $g(y) \geq 0$, satisfying

$$0 < \int_{-\infty}^{\infty}\left(\frac{g(y)}{e^{\mu_1 by}}\right)^q dy < \infty$$

we have the following inequality:

$$\left[\int_{-\infty}^{\infty} e^{q\sigma ax}\left(\int_{-\infty}^{\infty} k_\lambda(e^{ax}, e^{by})g(y)dy\right)^q dx\right]^{\frac{1}{q}}$$

$$> M\left[\int_{-\infty}^{\infty}\left(\frac{g(y)}{e^{\mu_1 by}}\right)^q dy\right]^{\frac{1}{p}}. \tag{3.54}$$

(iii) For any $f(x) \geq 0$, satisfying

$$0 < \int_{-\infty}^{\infty}\left(\frac{f(x)}{e^{\sigma ax}}\right)^p dx < \infty,$$

and $g(y) \geq 0$, satisfying

$$0 < \int_{-\infty}^{\infty}\left(\frac{g(y)}{e^{\mu_1 by}}\right)^q dy < \infty,$$

we have the following inequality:

$$\int_{-\infty}^{\infty}\int_{-\infty}^{\infty} k_\lambda(e^{ax}, e^{by})f(x)g(y)dxdy$$

$$> M\left[\int_{-\infty}^{\infty}\left(\frac{f(x)}{e^{\sigma ax}}\right)^p dx\right]^{\frac{1}{p}}\left[\int_{-\infty}^{\infty}\left(\frac{g(y)}{e^{\mu_1 by}}\right)^q dy\right]^{\frac{1}{q}}. \tag{3.55}$$

(iv) For

$$k_\lambda(\sigma) = \int_0^{\infty} k_\lambda(u, 1)u^{\sigma-1}du < \infty,$$

if there exists a constant $\delta_0 > 0$, such that

$$k_\lambda(\sigma \pm \delta_0) < \infty,$$

then we have

$$\mu_1 = \mu \quad and \quad \frac{k_\lambda(\sigma)}{|a|^{1/q}|b|^{1/p}} \geq M \ (> 0).$$

For $\mu_1 = \mu$, we obtain the following corollary:

Corollary 3.18 *If $0 < k_\lambda(\sigma) < \infty$, then the following statements (i), (ii) and (iii) are equivalent:*
(i) For any $f(x) \geq 0$, satisfying

$$0 < \int_{-\infty}^{\infty}\left(\frac{f(x)}{e^{\sigma ax}}\right)^p dx < \infty,$$

we have the following inequality:

$$\left[\int_{-\infty}^{\infty} e^{p\mu by} \left(\int_{-\infty}^{\infty} k_\lambda(e^{ax}, e^{by}) f(x) dx\right)^p dy\right]^{\frac{1}{p}}$$
$$> \frac{k_\lambda(\sigma)}{|a|^{1/q}|b|^{1/p}} \left[\int_{-\infty}^{\infty} \left(\frac{f(x)}{e^{\sigma ax}}\right)^p dx\right]^{\frac{1}{p}}. \tag{3.56}$$

(ii) For any $g(y) \geq 0$, satisfying

$$0 < \int_{-\infty}^{\infty} \left(\frac{g(y)}{e^{\mu by}}\right)^q dy < \infty$$

we have the following inequality:

$$\left[\int_{-\infty}^{\infty} e^{q\sigma ax} \left(\int_{-\infty}^{\infty} k_\lambda(e^{ax}, e^{by}) g(y) dy\right)^q dx\right]^{\frac{1}{q}}$$
$$> \frac{k_\lambda(\sigma)}{|a|^{1/q}|b|^{1/p}} \left[\int_{-\infty}^{\infty} \left(\frac{g(y)}{e^{\mu by}}\right)^q dy\right]^{\frac{1}{p}}. \tag{3.57}$$

(iii) For any $f(x) \geq 0$, satisfying

$$0 < \int_{-\infty}^{\infty} \left(\frac{f(x)}{e^{\sigma ax}}\right)^p dx < \infty,$$

and $g(y) \geq 0$, satisfying

$$0 < \int_{-\infty}^{\infty} \left(\frac{g(y)}{e^{\mu by}}\right)^q dy < \infty,$$

we have the following inequality:

$$\int_{-\infty}^{\infty} \int_{-\infty}^{\infty} k_\lambda(e^{ax}, e^{by}) f(x) g(y) dx dy$$
$$> \frac{k_\lambda(\sigma)}{|a|^{1/q}|b|^{1/p}} \left[\int_{-\infty}^{\infty} \left(\frac{f(x)}{e^{\sigma ax}}\right)^p dx\right]^{\frac{1}{p}} \left[\int_{-\infty}^{\infty} \left(\frac{g(y)}{e^{\mu by}}\right)^q dy\right]^{\frac{1}{q}}. \tag{3.58}$$

Moreover, if there exists a constant $\delta_0 > 0$, such that

$$k_\lambda(\sigma - \delta_0) < \infty \text{ or } k_\lambda(\sigma + \delta_0) < \infty,$$

then the constant factor

$$\frac{k_\lambda(\sigma)}{|a|^{1/q}|b|^{1/p}}$$

in (3.56), (3.57) and (3.58) is the best possible.

In particular, for

$$k_0(x, y) = e^{\frac{\alpha x}{y}} \csc h(\frac{x}{y}),$$

by Example 3.12, we have the following equivalent inequalities with the best possible constant factor

$$K_0(\sigma) := \frac{2^{1-\sigma}\Gamma(\sigma)}{|a|^{1/q}|b|^{1/p}} \zeta\left(\sigma, \frac{1-\alpha}{2}\right)$$

$(\alpha < 1, \sigma > 1)$:

$$\left[\int_{-\infty}^{\infty} e^{-p\sigma by} \left(\int_{-\infty}^{\infty} e^{\frac{\alpha x}{y}} \csc h(\frac{x}{y}) f(x) dx\right)^p dy\right]^{\frac{1}{p}}$$
$$> \frac{2^{1-\sigma}\Gamma(\sigma)}{|a|^{1/q}|b|^{1/p}} \zeta(\sigma, \frac{1-\alpha}{2}) \left[\int_{-\infty}^{\infty} \left(\frac{f(x)}{e^{\sigma ax}}\right)^p dx\right]^{\frac{1}{p}}, \tag{3.59}$$

$$\left[\int_{-\infty}^{\infty} e^{q\sigma ax} \left(\int_{-\infty}^{\infty} e^{\frac{\alpha x}{y}} \csc h(\frac{x}{y}) g(y) dy\right)^q dx\right]^{\frac{1}{q}}$$
$$> \frac{2^{1-\sigma}\Gamma(\sigma)}{|a|^{1/q}|b|^{1/p}} \zeta(\sigma, \frac{1-\alpha}{2}) \left[\int_{-\infty}^{\infty} \left(\frac{g(y)}{e^{-\sigma by}}\right)^q dy\right]^{\frac{1}{p}}, \tag{3.60}$$

$$\int_{-\infty}^{\infty} \int_{-\infty}^{\infty} e^{\frac{\alpha x}{y}} \csc h(\frac{x}{y}) f(x) g(y) dx dy$$
$$> \frac{2^{1-\sigma}\Gamma(\sigma)}{|a|^{1/q}|b|^{1/p}} \zeta(\sigma, \frac{1-\alpha}{2})$$
$$\times \left[\int_{-\infty}^{\infty} \left(\frac{f(x)}{e^{\sigma ax}}\right)^p dx\right]^{\frac{1}{p}} \left[\int_{-\infty}^{\infty} \left(\frac{g(y)}{e^{-\sigma by}}\right)^q dy\right]^{\frac{1}{q}}. \tag{3.61}$$

References

1. Kuang, J.C.: Applied Inequalities. Shangdong Science and Technology Press, Jinan, China (2004)
2. Kuang, J.C.: Real and Functional Analysis (continuation) (sec. vol.). Higher Education Press, Beijing, China (2015)

Chapter 4
Equivalent Statements of Two Kinds of Hardy-Type Integral Inequalities

In this chapter, we consider a few equivalent statements of two kinds of Hardy-type integral inequalities with a nonhomogeneous kernel related to certain parameters. Two kinds of Hardy-type integral inequalities with a homogeneous kernel are deduced. We also consider the operator expressions, some corollaries and a few particular examples involving the Hurwitz zeta function in the form of applications, as well as two kinds of Hardy-type integral inequalities in the whole plane.

4.1 Lemmas

Throughout this chapter we shall assume that: $p > 1$, $\frac{1}{p} + \frac{1}{q} = 1$, $\sigma_1, \sigma, \mu \in \mathbf{R}$, $\sigma + \mu = \lambda$, and $h(u)$ is a nonnegative measurable function in $(0, \infty)$, such that

$$k_1(\sigma) = \int_0^1 h(u)u^{\sigma-1}du \ (\geq 0),$$

$$k_2(\sigma) = \int_1^\infty h(u)u^{\sigma-1}du \ (\geq 0). \tag{4.1}$$

Lemma 4.1 *If $k_1(\sigma) > 0$, and if there exists a constant M_1 such that for any nonnegative measurable functions $f(x)$ and $g(y)$ in $(0, \infty)$, the following inequality*

$$\int_0^\infty g(y) \left(\int_0^{\frac{1}{y}} h(xy)f(x)dx \right) dy$$

$$\leq M_1 \left[\int_0^\infty x^{p(1-\sigma)-1}f^p(x)dx \right]^{\frac{1}{p}} \left[\int_0^\infty y^{q(1-\sigma_1)-1}g^q(y)dy \right]^{\frac{1}{q}} \tag{4.2}$$

© The Author(s), under exclusive licence to Springer Nature Switzerland AG 2019
B. Yang and M. Th. Rassias, *On Hilbert-Type and Hardy-Type Integral Inequalities and Applications*, SpringerBriefs in Mathematics,
https://doi.org/10.1007/978-3-030-29268-3_4

holds true, then we have

$$\sigma_1 = \sigma \text{ and } k_1(\sigma) \le M_1 \ (< \infty).$$

Proof Since $k_1(\sigma) > 0$, it follows that $h(u) > 0$ a.e. in an interval $I \ (\subset (0, 1))$.
If $\sigma_1 > \sigma$, then for $n \ge \frac{1}{\sigma_1 - \sigma} \ (n \in \mathbf{N})$, we set the following functions:

$$f_n(x) = \begin{cases} x^{\sigma + \frac{1}{pn} - 1}, \ 0 < x \le 1 \\ 0, \ x > 1 \end{cases},$$

$$g_n(y) = \begin{cases} 0, \ 0 < y < 1 \\ y^{\sigma_1 - \frac{1}{qn} - 1}, \ y \ge 1 \end{cases}.$$

We get

$$J_1 = \left[\int_0^\infty x^{p(1-\sigma)-1} f_n^p(x) dx \right]^{\frac{1}{p}} \left[\int_0^\infty y^{q(1-\sigma_1)-1} g_n^q(y) dy \right]^{\frac{1}{q}}$$

$$= \left[\int_0^1 x^{p(1-\sigma)-1} x^{p(\sigma + \frac{1}{pn} - 1)} dx \right]^{\frac{1}{p}}$$

$$\times \left[\int_1^\infty y^{q(1-\sigma_1)-1} y^{q(\sigma_1 - \frac{1}{qn} - 1)} dy \right]^{\frac{1}{q}}$$

$$= \left(\int_0^1 x^{\frac{1}{n}-1} dx \right)^{\frac{1}{p}} \left(\int_1^\infty y^{-\frac{1}{n}-1} dy \right)^{\frac{1}{q}} = n.$$

For fixed $y > 0$, setting $u = xy$, we obtain

$$I_1 := \int_0^\infty g_n(y) \left(\int_0^{\frac{1}{y}} h(xy) f_n(x) dx \right) dy$$

$$= \int_1^\infty \left(\int_0^{\frac{1}{y}} h(xy) x^{\sigma + \frac{1}{pn} - 1} dx \right) y^{\sigma_1 - \frac{1}{qn} - 1} dy$$

$$= \int_1^\infty y^{(\sigma_1 - \sigma) - \frac{1}{n} - 1} dy \int_0^1 h(u) u^{\sigma + \frac{1}{pn} - 1} du,$$

and then by (4.2) we have

$$\int_1^\infty y^{(\sigma_1 - \sigma) - \frac{1}{n} - 1} dy \int_0^1 h(u) u^{\sigma + \frac{1}{pn} - 1} du$$

$$= I_1 \le M_1 J_1 = M_1 n < \infty. \tag{4.3}$$

Since $(\sigma_1 - \sigma) - \frac{1}{n} \geq 0$, it follows that

$$\int_1^\infty y^{(\sigma_1-\sigma)-\frac{1}{n}-1}dy = \infty.$$

By (4.3), since

$$h(u)u^{\sigma+\frac{1}{pn}-1} > 0 \text{ a.e. in an interval } I \subset (0,1),$$

we obtain that

$$\int_0^1 h(u)u^{\sigma+\frac{1}{pn}-1}du > 0$$

and therefore

$$\infty \leq M_1 n < \infty,$$

which is a contradiction.

If $\sigma_1 < \sigma$, then for $n \geq \frac{1}{\sigma-\sigma_1}$ $(n \in \mathbf{N})$, we set the following functions:

$$\tilde{f}_n(x) = \begin{cases} 0, & 0 < x < 1 \\ x^{\sigma-\frac{1}{pn}-1}, & x \geq 1 \end{cases},$$

$$\tilde{g}_n(y) = \begin{cases} y^{\sigma_1+\frac{1}{qn}-1}, & 0 < y \leq 1 \\ 0, & y > 1 \end{cases},$$

and deduce that

$$\tilde{J}_1 = \left[\int_0^\infty x^{p(1-\sigma)-1}\tilde{f}_n^p(x)dx\right]^{\frac{1}{p}} \left[\int_0^\infty y^{q(1-\sigma_1)-1}\tilde{g}_n^q(y)dy\right]^{\frac{1}{q}}$$

$$= \left[\int_1^\infty x^{p(1-\sigma)-1}x^{p(\sigma-\frac{1}{pn}-1)}dx\right]^{\frac{1}{p}}$$

$$\times \left[\int_0^1 y^{q(1-\sigma_1)-1}y^{q(\sigma_1+\frac{1}{qn}-1)}dy\right]^{\frac{1}{q}}$$

$$= \left(\int_1^\infty x^{-\frac{1}{n}-1}dx\right)^{\frac{1}{p}} \left(\int_0^1 y^{\frac{1}{n}-1}dy\right)^{\frac{1}{q}} = n.$$

For fixed $x > 0$, setting $u = xy$, we obtain

$$\tilde{I}_1 := \int_0^\infty \tilde{f}_n(x) \left(\int_0^{\frac{1}{x}} h(xy)\tilde{g}_n(y)dy\right) dx$$

$$= \int_1^\infty \left(\int_0^{\frac{1}{x}} h(xy)y^{\sigma_1+\frac{1}{qn}-1}dy\right) x^{\sigma-\frac{1}{pn}-1}dx$$

$$= \int_1^\infty x^{(\sigma-\sigma_1)-\frac{1}{n}-1} dx \int_0^1 h(u) u^{\sigma_1+\frac{1}{qn}-1} du,$$

and then by Fubini's theorem (cf. [1]) and (4.2), we get that

$$\int_1^\infty x^{(\sigma-\sigma_1)-\frac{1}{n}-1} dx \int_0^1 h(u) u^{\sigma_1+\frac{1}{qn}-1} du$$

$$= \tilde{I}_1 = \int_0^\infty \tilde{g}_n(y) \left(\int_0^{\frac{1}{y}} h(xy) \tilde{f}_n(x) dx \right) dy$$

$$\leq M_1 \tilde{J}_1 = M_1 n. \tag{4.4}$$

Since $(\sigma - \sigma_1) - \frac{1}{n} \geq 0$, it follows that

$$\int_1^\infty x^{(\sigma-\sigma_1)-\frac{1}{n}-1} dx = \infty.$$

By (4.4), since

$$h(u) u^{\sigma_1+\frac{1}{qn}-1} > 0 \text{ a.e. in an interval } I \subset (0,1),$$

it follows that

$$\int_0^1 h(u) u^{\sigma_1+\frac{1}{qn}-1} du > 0,$$

and thus

$$\infty \leq M_1 n < \infty,$$

which is a contradiction.

Hence, we conclude that $\sigma_1 = \sigma$.

For $\sigma_1 = \sigma$, we reduce (4.4) as follows:

$$\int_0^1 h(u) u^{\sigma+\frac{1}{qn}-1} du \leq M_1. \tag{4.5}$$

Since $\{h(u) u^{\sigma+\frac{1}{qn}-1}\}_{n=1}^\infty$ is nonnegative and increasing in $(0, 1]$, by Levi's theorem (cf. [1]), we have

$$k_1(\sigma) = \int_0^1 \lim_{n\to\infty} h(u) u^{\sigma+\frac{1}{qn}-1} du$$

$$= \lim_{n\to\infty} \int_0^1 h(u) u^{\sigma+\frac{1}{qn}-1} du \leq M_1 (< \infty).$$

This completes the proof of the lemma.   □

Lemma 4.2 *If $k_2(\sigma) > 0$, and if there exists a constant M_2 such that for any non-negative measurable functions $f(x)$ and $g(y)$ in $(0, \infty)$, the following inequality*

$$\int_0^\infty g(y) \left(\int_{\frac{1}{y}}^\infty h(xy)f(x)dx \right) dy$$

$$\leq M_2 \left[\int_0^\infty x^{p(1-\sigma)-1} f^p(x)dx \right]^{\frac{1}{p}} \left[\int_0^\infty y^{q(1-\sigma_1)-1} g^q(y)dy \right]^{\frac{1}{q}} \qquad (4.6)$$

holds true, then we have

$$\sigma_1 = \sigma \quad and \quad k_2(\sigma) \leq M_2 \ (< \infty).$$

Proof Since $k_2(\sigma) > 0$, it follows that $h(u) > 0$ a.e. in an interval I_1 $(\subset (1, \infty))$.

If $\sigma_1 < \sigma$, then for $n \geq \frac{1}{\sigma - \sigma_1}$ $(n \in \mathbf{N})$, we set two functions $\tilde{f}_n(x)$ and $\tilde{g}_n(y)$, as in Lemma 4.1, and find

$$\tilde{J}_1 = \left[\int_0^\infty x^{p(1-\sigma)-1} \tilde{f}_n^p(x)dx \right]^{\frac{1}{p}} \left[\int_0^\infty y^{q(1-\sigma_1)-1} \tilde{g}_n^q(y)dy \right]^{\frac{1}{q}} = n.$$

Fix $y > 0$, setting $u = xy$, we obtain

$$\tilde{I}_2 := \int_0^\infty \tilde{g}_n(y) \left(\int_{\frac{1}{y}}^\infty h(xy) \tilde{f}_n(x)dx \right) dy$$

$$= \int_0^1 \left(\int_{\frac{1}{y}}^\infty h(xy)x^{\sigma - \frac{1}{pn} - 1}dx \right) y^{\sigma_1 + \frac{1}{qn} - 1}dy$$

$$= \int_0^1 y^{(\sigma_1 - \sigma) + \frac{1}{n} - 1}dy \int_1^\infty h(u)u^{\sigma - \frac{1}{pn} - 1}du,$$

and then by (4.6) we have

$$\int_0^1 y^{(\sigma_1 - \sigma) + \frac{1}{n} - 1}dy \int_1^\infty h(u)u^{\sigma - \frac{1}{pn} - 1}du$$

$$= \tilde{I}_2 \leq M_2 \tilde{J}_1 = M_2 n < \infty. \qquad (4.7)$$

Since $(\sigma_1 - \sigma) + \frac{1}{n} \leq 0$, it follows that

$$\int_0^1 y^{(\sigma_1 - \sigma) + \frac{1}{n} - 1}dy = \infty.$$

By (4.7), since

$$h(u)u^{\sigma - \frac{1}{pn} - 1} > 0 \quad \text{a.e. in an interval } I_1 \subset (1, \infty),$$

we find that

$$\int_1^\infty h(u)u^{\sigma - \frac{1}{pn} - 1} du > 0,$$

and therefore

$$\infty \le M_2 n < \infty,$$

which is a contradiction.

If $\sigma_1 > \sigma$, then for $n \ge \frac{1}{\sigma_1 - \sigma}$ $(n \in \mathbf{N})$, we set two functions $f_n(x)$ and $g_n(y)$ as in Lemma 4.1 and find

$$J_1 = \left[\int_0^\infty x^{p(1-\sigma)-1} f_n^p(x) dx \right]^{\frac{1}{p}} \left[\int_0^\infty y^{q(1-\sigma_1)-1} g_n^q(y) dy \right]^{\frac{1}{q}} = n.$$

For fixed $x > 0$, setting $u = xy$, we obtain

$$I_2 := \int_0^\infty f_n(x) \left(\int_{\frac{1}{x}}^\infty h(xy) g_n(y) dy \right) dx$$

$$= \int_0^1 \left(\int_{\frac{1}{x}}^\infty h(xy) y^{\sigma_1 - \frac{1}{qn} - 1} dy \right) x^{\sigma + \frac{1}{pn} - 1} dx$$

$$= \int_0^1 x^{(\sigma - \sigma_1) + \frac{1}{n} - 1} dx \int_1^\infty h(u) u^{\sigma_1 - \frac{1}{qn} - 1} du,$$

and then by Fubini's theorem (cf. [1]) and (4.6), we obtain that

$$\int_0^1 x^{(\sigma - \sigma_1) + \frac{1}{n} - 1} dx \int_1^\infty h(u) u^{\sigma_1 - \frac{1}{qn} - 1} du$$

$$= I_2 = \int_0^\infty g_n(y) \left(\int_{\frac{1}{y}}^\infty h(xy) f_n(x) dx \right) dy$$

$$\le M_2 J_1 = M_2 n < \infty. \tag{4.8}$$

Since $(\sigma - \sigma_1) + \frac{1}{n} \le 0$, it follows that

$$\int_0^1 x^{(\sigma - \sigma_1) + \frac{1}{n} - 1} dx = \infty.$$

By (4.8), in view of the fact that

$$h(u) u^{\sigma_1 - \frac{1}{qn} - 1} > 0 \text{ a.e. in an interval } I_1 \subset (1, \infty),$$

we derive that

$$\int_1^\infty h(u) u^{\sigma_1 - \frac{1}{qn} - 1} du > 0,$$

and hence

$$\infty \le M_2 n < \infty,$$

which is a contradiction.

Hence, we conclude the fact that $\sigma_1 = \sigma$.

For $\sigma_1 = \sigma$, we reduce (4.8) as follows:

$$\int_1^\infty h(u) u^{\sigma - \frac{1}{qn} - 1} du \le M_2. \tag{4.9}$$

Since $\{h(u) u^{\sigma - \frac{1}{qn} - 1}\}_{n=1}^\infty$ is nonnegative and increasing in $[1, \infty)$, applying again Levi's theorem (cf. [1]), we obtain that

$$k_2(\sigma) = \int_1^\infty \lim_{n \to \infty} h(u) u^{\sigma - \frac{1}{qn} - 1} du$$

$$= \lim_{n \to \infty} \int_1^\infty h(u) u^{\sigma - \frac{1}{qn} - 1} du \le M_2(< \infty).$$

This completes the proof of the lemma. $\square$

4.2 Hardy-Type Integral Inequalities of the First Kind

Theorem 4.3 *Let M_1 be a constant. If $k_1(\sigma) > 0$, then the following statements (i), (ii) and (iii) are equivalent: (i) For any $f(x) \ge 0$, satisfying*

$$0 < \int_0^\infty x^{p(1-\sigma)-1} f^p(x) dx < \infty,$$

we have the following Hardy-type integral inequality of the first kind with a nonhomogeneous kernel:

$$J := \left[\int_0^\infty y^{p\sigma_1 - 1} \left(\int_0^{\frac{1}{y}} h(xy) f(x) dx \right)^p dy \right]^{\frac{1}{p}}$$

$$< M_1 \left[\int_0^\infty x^{p(1-\sigma)-1} f^p(x) dx \right]^{\frac{1}{p}}. \tag{4.10}$$

(ii) For any $f(x) \ge 0$, satisfying

$$0 < \int_0^\infty x^{p(1-\sigma)-1} f^p(x) dx < \infty,$$

and $g(y) \geq 0$, satisfying

$$0 < \int_0^\infty y^{q(1-\sigma_1)-1} g^q(y) dy < \infty,$$

we have the following inequality:

$$I := \int_0^\infty g(y) \left(\int_0^{\frac{1}{y}} h(xy) f(x) dx \right) dy$$

$$< M_1 \left[\int_0^\infty x^{p(1-\sigma)-1} f^p(x) dx \right]^{\frac{1}{p}} \left[\int_0^\infty y^{q(1-\sigma_1)-1} g^q(y) dy \right]^{\frac{1}{q}}. \quad (4.11)$$

(iii) $\sigma_1 = \sigma$, and $k_1(\sigma) \leq M_1 < \infty$.

If statement (iii) holds true, then the constant $M_1 = k_1(\sigma)$ $(\in \mathbf{R}_+)$ in (4.10) and (4.11) is the best possible.

Proof $(i) \Rightarrow (ii)$. By Hölder's inequality (cf. [2]), we have

$$I = \int_0^\infty \left(y^{\sigma_1 - \frac{1}{p}} \int_0^{\frac{1}{y}} h(xy) f(x) dx \right) \left(y^{\frac{1}{p} - \sigma_1} g(y) \right) dy$$

$$\leq J \left[\int_0^\infty y^{q(1-\sigma_1)-1} g^q(y) dy \right]^{\frac{1}{q}}. \quad (4.12)$$

Then by (4.10), we have (4.11).

$(ii) \Rightarrow (iii)$. By Lemma 4.1, we have $\sigma_1 = \sigma$, and $k_1(\sigma) \leq M_1 < \infty$.

$(iii) \Rightarrow (i)$. For fixed $y > 0$, setting $u = xy$, we obtain the following weight function:

$$\omega_1(\sigma, y) := y^\sigma \int_0^{\frac{1}{y}} h(xy) x^{\sigma-1} dx$$

$$= \int_0^1 h(u) u^{\sigma-1} du = k_1(\sigma) \ (y > 0). \quad (4.13)$$

By Hölder's inequality with weight and (4.13), for $y \in (0, \infty)$, we have

$$\left(\int_0^{\frac{1}{y}} h(xy) f(x) dx \right)^p$$

$$= \left\{ \int_0^{\frac{1}{y}} h(xy) \left[\frac{y^{(\sigma-1)/p}}{x^{(\sigma-1)/q}} f(x) \right] \left[\frac{x^{(\sigma-1)/q}}{y^{(\sigma-1)/p}} \right] dx \right\}^p$$

$$\leq \int_0^{\frac{1}{y}} h(xy) \frac{y^{\sigma-1} f^p(x)}{x^{(\sigma-1)p/q}} dx \left[\int_0^{\frac{1}{y}} h(xy) \frac{x^{\sigma-1}}{y^{(\sigma-1)q/p}} dx \right]^{p-1}$$

$$= \left[\omega_1(\sigma, y) y^{q(1-\sigma)-1} \right]^{p-1} \int_0^{\frac{1}{y}} h(xy) \frac{y^{\sigma-1}}{x^{(\sigma-1)p/q}} f^p(x) dx$$

$$= (k_1(\sigma))^{p-1} y^{-p\sigma+1} \int_0^{\frac{1}{y}} h(xy) \frac{y^{\sigma-1}}{x^{(\sigma-1)p/q}} f^p(x) dx. \tag{4.14}$$

If (4.14) takes the form of equality for some $y \in (0, \infty)$, then (cf. [2]) there exist constants A and B, such that they are not all zero, and

$$A \frac{y^{\sigma-1}}{x^{(\sigma-1)p/q}} f^p(x) = B \frac{x^{\sigma-1}}{y^{(\sigma-1)q/p}} \quad a.e. \text{ in } \mathbf{R}_+.$$

Let us suppose that $A \neq 0$ (otherwise $B = A = 0$). It follows that

$$x^{p(1-\sigma)-1} f^p(x) = y^{q(1-\sigma)} \frac{B}{Ax} \quad a.e. \text{ in } \mathbf{R}_+,$$

which contradicts the fact that

$$0 < \int_0^\infty x^{p(1-\sigma)-1} f^p(x) dx < \infty.$$

Hence, (4.14) takes the form of strict inequality.

For $\sigma_1 = \sigma$, by Fubini's theorem (cf. [1]), we have

$$J < (k_1(\sigma))^{\frac{1}{q}} \left\{ \int_0^\infty \left[\int_0^{\frac{1}{y}} h(xy) \frac{y^{\sigma-1}}{x^{(\sigma-1)p/q}} f^p(x) dx \right] dy \right\}^{\frac{1}{p}}$$

$$= (k_1(\sigma))^{\frac{1}{q}} \left\{ \int_0^\infty \left[\int_0^{\frac{1}{x}} h(xy) \frac{y^{\sigma-1}}{x^{(\sigma-1)(p-1)}} dy \right] f^p(x) dx \right\}^{\frac{1}{p}}$$

$$= (k_1(\sigma))^{\frac{1}{q}} \left[\int_0^\infty \omega_1(\sigma, x) x^{p(1-\sigma)-1} f^p(x) dx \right]^{\frac{1}{p}}$$

$$= k_1(\sigma) \left[\int_0^\infty x^{p(1-\sigma)-1} f^p(x) dx \right]^{\frac{1}{p}}.$$

Since $k_1(\sigma) \leq M_1$, (4.10) follows.

Therefore, statements (i), (ii) and (iii) are equivalent.

When statement (iii) holds true, if there exists a constant $M_1 \leq k_1(\sigma)$, such that (4.11) is valid, then in view of the assumption, we have $k_1(\sigma) \leq M_1$. Hence, $M_1 = k_1(\sigma)$ in (4.11) is the best possible.

The constant factor $M_1 = k_1(\sigma)$ ($\in \mathbf{R}_+$) in (4.10) is still the best possible. Otherwise, by (4.12) (for $\sigma_1 = \sigma$), we would conclude that the constant factor $M_1 = k_1(\sigma)$ in (4.11) is not the best possible.

This completes the proof of the theorem. □

Remark 4.4 If $k_1(\sigma) = 0$, then $h(u) = 0$ a.e. in $(0, 1]$, and we can still show that statement (i) is equivalent to statement (ii). But statement (ii) does not deduce $\sigma_1 = \sigma$ of statement (iii).

In particular, for $\sigma = \sigma_1 = \frac{1}{p}$ in Theorem 4.3, we obtain the following corollary:

Corollary 4.5 *Let M_1 be a constant. If $k_1(\frac{1}{p}) > 0$, then the following statements (i), (ii) and (iii) are equivalent:*
(i) For any $f(x) \geq 0$, satisfying

$$0 < \int_0^\infty x^{p-2} f^p(x) dx < \infty,$$

we have the following inequality:

$$\left[\int_0^\infty \left(\int_0^{\frac{1}{y}} h(xy) f(x) dx \right)^p dy \right]^{\frac{1}{p}} < M_1 \left(\int_0^\infty x^{p-2} f^p(x) dx \right)^{\frac{1}{p}}. \qquad (4.15)$$

(ii) For any $f(x) \geq 0$, satisfying

$$0 < \int_0^\infty x^{p-2} f^p(x) dx < \infty,$$

and $g(y) \geq 0$, satisfying

$$0 < \int_0^\infty g^q(y) dy < \infty,$$

we have the following inequality:

$$\int_0^\infty g(y) \left(\int_0^{\frac{1}{y}} h(xy) f(x) dx \right) dy$$

$$< M_1 \left(\int_0^\infty x^{p-2} f^p(x) dx \right)^{\frac{1}{p}} \left(\int_0^\infty g^q(y) dy \right)^{\frac{1}{q}}. \qquad (4.16)$$

(iii) $k_1(\frac{1}{p}) \leq M_1 < \infty$.
If statement (iii) holds true, then the constant $M_1 = k_1(\frac{1}{p}) (\in \mathbf{R}_+)$ in (4.15) and (4.16) is the best possible.

Setting

$$y = \frac{1}{Y}, \quad G(Y) = g\left(\frac{1}{Y}\right) \frac{1}{Y^2}$$

in Theorem 4.3, and then replacing Y by y, we get:

Corollary 4.6 *Let M_1 be a constant. If $k_1(\sigma) > 0$, then the following statements (i), (ii) and (iii) are equivalent:*

(i) For any $f(x) \geq 0$, satisfying

$$0 < \int_0^\infty x^{p(1-\sigma)-1} f^p(x)dx < \infty,$$

we have the following inequality:

$$\left[\int_0^\infty y^{-p\sigma_1-1} \left(\int_0^y h\left(\frac{x}{y}\right) f(x)dx\right)^p dy\right]^{\frac{1}{p}}$$

$$< M_1 \left[\int_0^\infty x^{p(1-\sigma)-1} f^p(x)dx\right]^{\frac{1}{p}}. \tag{4.17}$$

(ii) For any $f(x) \geq 0$, satisfying

$$0 < \int_0^\infty x^{p(1-\sigma)-1} f^p(x)dx < \infty,$$

and $G(y) \geq 0$, satisfying

$$0 < \int_0^\infty y^{q(1+\sigma_1)-1} G^q(y)dy < \infty,$$

we have the following inequality:

$$\int_0^\infty G(y) \left(\int_0^y h\left(\frac{x}{y}\right) f(x)dx\right) dy$$

$$< M_1 \left[\int_0^\infty x^{p(1-\sigma)-1} f^p(x)dx\right]^{\frac{1}{p}} \left[\int_0^\infty y^{q(1+\sigma_1)-1} G^q(y)dy\right]^{\frac{1}{q}}. \tag{4.18}$$

(iii) $\sigma_1 = \sigma$, and $k_1(\sigma) \leq M_1 < \infty$.

If statement (iii) holds true, then the constant $M_1 = k_1(\sigma) (\in \mathbf{R}_+)$ in (4.17) and (4.18) is the best possible.

Note. *$h(\frac{x}{y})$ is a homogeneous function of degree 0, namely,*

$$h\left(\frac{x}{y}\right) = k_0(x, y).$$

Setting

$$h(u) = k_\lambda(u, 1),$$

where $k_\lambda(x, y)$ is a homogeneous function of degree $-\lambda$ ($\in \mathbf{R}$), for $g(y) = y^\lambda G(y)$ and $\mu_1 = \lambda - \sigma_1$ in Corollary 4.6, we have the following:

Corollary 4.7 *Let M_1 be a constant. If*

$$k_\lambda^{(1)}(\sigma) = \int_0^1 k_\lambda(u, 1) u^{\sigma-1} du > 0,$$

then the following statements (i), (ii) and (iii) are equivalent:
  (i) For any $f(x) \geq 0$, satisfying

$$0 < \int_0^\infty x^{p(1-\sigma)-1} f^p(x) dx < \infty,$$

we have the following Hardy-type integral inequality of the first kind with the homogeneous kernel:

$$\left[\int_0^\infty y^{p\mu_1-1} \left(\int_0^y k_\lambda(x, y) f(x) dx \right)^p dy \right]^{\frac{1}{p}}$$
$$< M_1 \left[\int_0^\infty x^{p(1-\sigma)-1} f^p(x) dx \right]^{\frac{1}{p}}. \tag{4.19}$$

  (ii) For any $f(x) \geq 0$, satisfying

$$0 < \int_0^\infty x^{p(1-\sigma)-1} f^p(x) dx < \infty,$$

and $g(y) \geq 0$, satisfying

$$0 < \int_0^\infty y^{q(1-\mu_1)-1} g^q(y) dy < \infty,$$

we have the following inequality:

$$\int_0^\infty g(y) \left(\int_0^y k_\lambda(x, y) f(x) dx \right) dy$$
$$< M_1 \left[\int_0^\infty x^{p(1-\sigma)-1} f^p(x) dx \right]^{\frac{1}{p}} \left[\int_0^\infty y^{q(1-\mu_1)-1} g^q(y) dy \right]^{\frac{1}{q}}. \tag{4.20}$$

  (iii) $\mu_1 = \mu$, and $k_\lambda^{(1)}(\sigma) \leq M_1 < \infty$.
  If statement (iii) holds, then the constant $M_1 = k_\lambda^{(1)}(\sigma)$ ($\in \mathbf{R}_+$) in (4.19) and (4.20) is the best possible.

Remark 4.8 If $h(u) = k_\lambda(u, 1)$, and $k_1(\sigma) = k_\lambda^{(1)}(\sigma) > 0$, then Theorem 4.3 and Corollary 4.7 are equivalent.

In particular, for $\lambda = 1, \sigma = \frac{1}{q}, \mu_1 = \mu = \frac{1}{p}$ in Corollary 4.7, we have

Corollary 4.9 *Let M_1 be a constant. If $k_1^{(1)}(\frac{1}{q}) > 0$, then the following statements (i), (ii) and (iii) are equivalent:*

(i) For any $f(x) \geq 0$, satisfying

$$0 < \int_0^\infty f^p(x)dx < \infty,$$

we have the following inequality:

$$\left[\int_0^\infty \left(\int_0^y k_1(x, y)f(x)dx \right)^p dy \right]^{\frac{1}{p}} < M_1 \left(\int_0^\infty f^p(x)dx \right)^{\frac{1}{p}}. \tag{4.21}$$

(ii) For any $f(x) \geq 0$, satisfying

$$0 < \int_0^\infty f^p(x)dx < \infty,$$

and $g(y) \geq 0$, satisfying

$$0 < \int_0^\infty g^q(y)dy < \infty,$$

we have the following inequality:

$$I = \int_0^\infty g(y) \left(\int_0^y k_1(x, y)f(x)dx \right) dy$$

$$< M_1 \left(\int_0^\infty f^p(x)dx \right)^{\frac{1}{p}} \left(\int_0^\infty g^q(y)dy \right)^{\frac{1}{q}}. \tag{4.22}$$

(iii) $k_1^{(1)}(\frac{1}{q}) \leq M_1 \ (< \infty)$.

If statement (iii) holds true, then the constant $M_1 = k_1^{(1)}(\frac{1}{q}) \ (\in \mathbf{R}_+)$ in (4.21) and (4.22) is the best possible.

4.3 Hardy-Type Integral Inequalities of the Second Kind

Similarly, we can obtain the following weight function:

$$\omega_2(\sigma, y) := y^\sigma \int_{\frac{1}{y}}^\infty h(xy)x^{\sigma-1}dx$$

$$= \int_1^\infty h(u)u^{\sigma-1}du = k_2(\sigma) \ (y > 0),$$

and then in view of Lemma 4.2, in the same manner, we obtain the following theorem:

Theorem 4.10 *Let M_2 be a constant. If $k_2(\sigma) > 0$, then the following statements (i), (ii) and (iii) are equivalent:*

(i) For any $f(x) \geq 0$, satisfying

$$0 < \int_0^\infty x^{p(1-\sigma)-1} f^p(x) dx < \infty,$$

we have the following Hardy-type inequality of the second kind with the nonhomogeneous kernel:

$$\left[\int_0^\infty y^{p\sigma_1 - 1} \left(\int_{\frac{1}{y}}^\infty h(xy) f(x) dx \right)^p dy \right]^{\frac{1}{p}}$$

$$< M_2 \left[\int_0^\infty x^{p(1-\sigma)-1} f^p(x) dx \right]^{\frac{1}{p}}. \tag{4.23}$$

(ii) For any $f(x) \geq 0$, satisfying

$$0 < \int_0^\infty x^{p(1-\sigma)-1} f^p(x) dx < \infty,$$

and $g(y) \geq 0$, satisfying

$$0 < \int_0^\infty y^{q(1-\sigma_1)-1} g^q(y) dy < \infty,$$

we have the following inequality:

$$\int_0^\infty g(y) \left(\int_{\frac{1}{y}}^\infty h(xy) f(x) dx \right) dy$$

$$< M_2 \left[\int_0^\infty x^{p(1-\sigma)-1} f^p(x) dx \right]^{\frac{1}{p}} \left[\int_0^\infty y^{q(1-\sigma_1)-1} g^q(y) dy \right]^{\frac{1}{q}}. \tag{4.24}$$

(iii) $\sigma_1 = \sigma$, and $k_2(\sigma) \leq M_2 \ (< \infty)$.

If statement (iii) holds true, then the constant $M_2 = k_2(\sigma) \ (\in \mathbf{R}_+)$ in (4.23) and (4.24) is the best possible.

Remark 4.11 If $k_2(\sigma) = 0$, then $h(u) = 0$ a.e. in $(1, \infty)$, and we can still show that statement (i) is equivalent to statement (ii), but statement (ii) does not deduce $\sigma_1 = \sigma$ of statement (iii).

In particular, for $\sigma = \sigma_1 = \frac{1}{p}$ in Theorem 4.10, we obtain the following:

Corollary 4.12 *Let M_2 be a constant. If $k_2(\frac{1}{p}) > 0$, then the following statements (i), (ii) and (iii) are equivalent:*

(i) For any $f(x) \geq 0$, satisfying

$$0 < \int_0^\infty x^{p-2} f^p(x) dx < \infty,$$

we have the following inequality:

$$\left[\int_0^\infty \left(\int_{\frac{1}{y}}^\infty h(xy) f(x) dx \right)^p dy \right]^{\frac{1}{p}} < M_2 \left(\int_0^\infty x^{p-2} f^p(x) dx \right)^{\frac{1}{p}}. \qquad (4.25)$$

(ii) For any $f(x) \geq 0$, satisfying

$$0 < \int_0^\infty x^{p-2} f^p(x) dx < \infty,$$

and $g(y) \geq 0$, satisfying

$$0 < \int_0^\infty g^q(y) dy < \infty,$$

we have the following inequality:

$$\int_0^\infty g(y) \left(\int_{\frac{1}{y}}^\infty h(xy) f(x) dx \right) dy$$

$$< M_2 \left(\int_0^\infty x^{p-2} f^p(x) dx \right)^{\frac{1}{p}} \left(\int_0^\infty g^q(y) dy \right)^{\frac{1}{q}}. \qquad (4.26)$$

(iii) $k_2(\frac{1}{p}) \leq M_2 \ (< \infty)$.

If statement (iii) holds true, then the constant $M_2 = k_2(\frac{1}{p}) \ (\in \mathbf{R}_+)$ in (4.25) and (4.26) is the best possible.

Setting

$$y = \frac{1}{Y}, \quad G(Y) = g\left(\frac{1}{Y}\right) \frac{1}{Y^2}$$

in Theorem 4.10, and then replacing Y by y, we obtain the following:

Corollary 4.13 *Suppose that M_2 is a constant. If $k_2(\sigma) > 0$, then the following statements (i), (ii) and (iii) are equivalent:*

(i) For any $f(x) \geq 0$, satisfying

$$0 < \int_0^\infty x^{p(1-\sigma)-1} f^p(x) dx < \infty,$$

we have the following inequality:

$$\left[\int_0^\infty y^{-p\sigma_1-1} \left(\int_y^\infty h\left(\frac{x}{y}\right) f(x)dx \right)^p dy \right]^{\frac{1}{p}}$$
$$< M_2 \left[\int_0^\infty x^{p(1-\sigma)-1} f^p(x)dx \right]^{\frac{1}{p}}. \tag{4.27}$$

(ii) For any $f(x), G(y) \geq 0$, satisfying

$$0 < \int_0^\infty x^{p(1-\sigma)-1} f^p(x)dx < \infty,$$

and $G(y) \geq 0$, satisfying

$$0 < \int_0^\infty y^{q(1+\sigma_1)-1} G^q(y)dy < \infty,$$

we have the following inequality:

$$\int_0^\infty G(y) \left(\int_y^\infty h\left(\frac{x}{y}\right) f(x)dx \right) dy$$
$$< M_2 \left[\int_0^\infty x^{p(1-\sigma)-1} f^p(x)dx \right]^{\frac{1}{p}} \left[\int_0^\infty y^{q(1+\sigma_1)-1} G^q(y)dy \right]^{\frac{1}{q}}. \tag{4.28}$$

(iii) $\sigma_1 = \sigma$, and $k_2(\sigma) \leq M_2 (< \infty)$.
If statement (iii) holds true, then the constant $M_2 = k_2(\sigma) (\in \mathbf{R}_+)$ in (4.27) and (4.28) is the best possible.

Setting $h(u) = k_\lambda(u, 1)$, where, $k_\lambda(x, y)$ is a homogeneous function of degree $-\lambda (\in \mathbf{R})$, for $g(y) = y^\lambda G(y)$ and $\mu_1 = \lambda - \sigma_1$ in Corollary 4.13, we have:

Corollary 4.14 *Suppose that M_2 is a constant. If*

$$k_\lambda^{(2)}(\sigma) = \int_1^\infty k_\lambda(u, 1)u^{\sigma-1}du > 0,$$

then the following statements (i), (ii) and (iii) are equivalent:

(i) For any $f(x) \geq 0$, satisfying

$$0 < \int_0^\infty x^{p(1-\sigma)-1} f^p(x)dx < \infty,$$

we have the following second kind of Hardy-type integral inequality with the homogeneous kernel:

$$\left[\int_0^\infty y^{p\mu_1-1}\left(\int_y^\infty k_\lambda(x,y)f(x)dx\right)^p dy\right]^{\frac{1}{p}}$$

$$< M_2\left[\int_0^\infty x^{p(1-\sigma)-1}f^p(x)dx\right]^{\frac{1}{p}}. \tag{4.29}$$

(ii) For any $f(x) \geq 0$, satisfying

$$0 < \int_0^\infty x^{p(1-\sigma)-1}f^p(x)dx < \infty,$$

and $g(y) \geq 0$, satisfying

$$0 < \int_0^\infty y^{q(1-\mu_1)-1}g^q(y)dy < \infty,$$

we have the following inequality:

$$\int_0^\infty g(y)\left(\int_y^\infty k_\lambda(x,y)f(x)dx\right)dy$$

$$< M_2\left[\int_0^\infty x^{p(1-\sigma)-1}f^p(x)dx\right]^{\frac{1}{p}}\left[\int_0^\infty y^{q(1-\mu_1)-1}g^q(y)dy\right]^{\frac{1}{q}}. \tag{4.30}$$

(iii) $\mu_1 = \mu$, and $k_\lambda^{(2)}(\sigma) \leq M_2 (< \infty)$.

If statement (iii) holds true, then the constant $M_2 = k_\lambda^{(2)}(\sigma) (\in \mathbf{R}_+)$ in (4.29) and (4.30) is the best possible.

Remark 4.15 (a) If $k_\lambda^{(2)}(\sigma) = 0$, then $k_\lambda(u, 1) = 0$ a.e. in $(1, \infty)$, we still can show that statement (i) is equivalent to statement (ii), but statement (ii) does not deduce to $\sigma_1 = \sigma$ in statement (iii). (b) If $h(u) = k_\lambda(u, 1)$, and $k_2(\sigma) = k_\lambda^{(2)}(\sigma) > 0$, then Theorem 4.10 and Corollary 4.14 are equivalent.

In particular, for $\lambda = 1, \sigma = \frac{1}{q}, \mu_1 = \mu = \frac{1}{p}$ in Corollary 4.14, we have:

Corollary 4.16 *Let M_2 be a constant. If $k_1^{(2)}(\frac{1}{q}) > 0$, then the following statements (i), (ii) and (iii) are equivalent:*
(i) For any $f(x) \geq 0$, satisfying

$$0 < \int_0^\infty f^p(x)dx < \infty,$$

we have the following inequality:

$$\left[\int_0^\infty\left(\int_y^\infty k_1(x,y)f(x)dx\right)^p dy\right]^{\frac{1}{p}} < M_2\left(\int_0^\infty f^p(x)dx\right)^{\frac{1}{p}}. \tag{4.31}$$

(ii) For any $f(x) \geq 0$, satisfying

$$0 < \int_0^\infty f^p(x)dx < \infty,$$

and $g(y) \geq 0$, satisfying

$$0 < \int_0^\infty g^q(y)dy < \infty,$$

we have the following inequality:

$$I = \int_0^\infty g(y) \left(\int_y^\infty k_1(x, y)f(x)dx \right) dy$$

$$< M_2 \left(\int_0^\infty f^p(x)dx \right)^{\frac{1}{p}} \left(\int_0^\infty g^q(y)dy \right)^{\frac{1}{q}}. \tag{4.32}$$

(iii) $k_1^{(2)}(\frac{1}{q}) \leq M_2 \ (< \infty)$.

If statement (iii) holds true, then the constant $M_2 = k_1^{(2)}(\frac{1}{q}) \ (\in \mathbf{R}_+)$ in (4.31) and (4.32) is the best possible.

4.4 Operator Expressions and Some Examples

We set the following functions:

$$\varphi(x) = x^{p(1-\sigma)-1}, \ \ \psi(y) = y^{q(1-\sigma)-1}, \ \ \phi(y) = y^{q(1-\mu)-1},$$

wherefrom

$$\psi^{1-p}(y) = y^{p\sigma-1}, \ \ \phi^{1-p}(y) = y^{p\mu-1} \ \ (x, y \in \mathbf{R}_+).$$

We also consider the following real normed linear spaces:

$$L_{p,\varphi}(\mathbf{R}_+) = \left\{ f : \|f\|_{p,\varphi} = \left(\int_0^\infty \varphi(x)|f(x)|^p dx \right)^{\frac{1}{p}} < \infty \right\},$$

wherefrom

$$L_{q,\psi}(\mathbf{R}_+) = \left\{ g : \|g\|_{q,\psi} = \left(\int_0^\infty \psi(y)|g(y)|^q dy \right)^{\frac{1}{q}} < \infty \right\},$$

$$L_{q,\phi}(\mathbf{R}_+) = \left\{ g : \|g\|_{q,\phi} = \left(\int_0^\infty \phi(y)|g(y)|^q dy \right)^{\frac{1}{q}} < \infty \right\},$$

$$L_{p,\psi^{1-p}}(\mathbf{R}_+) = \left\{ h : ||h||_{p,\psi^{1-p}} = \left(\int_0^\infty \psi^{1-p}(y)|h(y)|^p dy \right)^{\frac{1}{p}} < \infty \right\},$$

$$L_{q,\phi^{1-p}}(\mathbf{R}_+) = \left\{ h : ||h||_{p,\phi^{1-p}} = \left(\int_0^\infty \phi^{1-p}(y)|h(y)|^p dy \right)^{\frac{1}{p}} < \infty \right\}.$$

(a) From Theorem 4.3 (for $\sigma_1 = \sigma$) and Remark 4.4, for $f \in L_{p,\varphi}(\mathbf{R}_+)$, setting

$$h_1(y) := \int_0^{\frac{1}{y}} h(xy)f(x)dx \ (y \in \mathbf{R}_+),$$

by (4.10), we have

$$||h_1||_{p,\psi^{1-p}} = \left[\int_0^\infty \psi^{1-p}(y)h_1^p(y)dy \right]^{\frac{1}{p}} < M_1||f||_{p,\varphi} < \infty. \tag{4.33}$$

Definition 4.17 We define a Hardy-type integral operator of the first kind with the nonhomogeneous kernel

$$T_1^{(1)} : L_{p,\varphi}(\mathbf{R}_+) \to L_{p,\psi^{1-p}}(\mathbf{R}_+)$$

as follows:
For any $f \in L_{p,\varphi}(\mathbf{R}_+)$, there exists a unique representation

$$T_1^{(1)}f = h_1 \in L_{p,\psi^{1-p}}(\mathbf{R}_+),$$

satisfying

$$T_1^{(1)}f(y) = h_1(y) ,$$

for any $y \in \mathbf{R}_+$.

In view of (4.33), it follows that

$$||T_1^{(1)}f||_{p,\psi^{1-p}} = ||h_1||_{p,\psi^{1-p}} \le M_1||f||_{p,\varphi},$$

and then the operator $T_1^{(1)}$ is bounded satisfying

$$||T_1^{(1)}|| = \sup_{f(\neq\theta)\in L_{p,\varphi}(\mathbf{R}_+)} \frac{||T_1^{(1)}f||_{p,\psi^{1-p}}}{||f||_{p,\varphi}} \le M_1.$$

If we define the formal inner product of $T_1^{(1)}f$ and g as follows:

$$(T_1^{(1)} f, g) = \int_0^\infty \left(\int_0^{\frac{1}{y}} h(xy) f(x) dx \right) g(y) dy,$$

we can then re-express Theorem 4.3 (for $\sigma_1 = \sigma$) by the use of Remark 4.4 as follows:

Theorem 4.18 *Let M_1 be a constant. The following statements (i), (ii) and (iii) are equivalent:*

(i) For any $f(x) \geq 0, f \in L_{p,\varphi}(\mathbf{R}_+)$, $||f||_{p,\varphi} > 0$, we have the following inequality:

$$||T_1^{(1)} f||_{p,\psi^{1-p}} < M_1 ||f||_{p,\varphi}. \tag{4.34}$$

(ii) For any $f(x), g(y) \geq 0, f \in L_{p,\varphi}(\mathbf{R}_+), g \in L_{q,\psi}(\mathbf{R}_+)$, $||f||_{p,\varphi}, ||g||_{q,\psi} > 0$, we have the following inequality:

$$(T_1^{(1)} f, g) < M_1 ||f||_{p,\varphi} ||g||_{q,\psi}. \tag{4.35}$$

(iii) $k_1(\sigma) \leq M_1 (< \infty)$.
If statement (iii) holds true, then we have

$$||T_1^{(1)}|| = k_1(\sigma) \leq M_1.$$

Note. If $k_1(\sigma) = 0$, then $T_1^{(1)} = \theta$ and $||T_1^{(1)}|| = 0 = k_1(\sigma) \leq M_1$.
(b) In view of Corollary 4.12 (with $\mu_1 = \mu$), for $f \in L_{p,\varphi}(\mathbf{R}_+)$, setting

$$h_2(y) := \int_0^y k_\lambda(x, y) f(x) dx \ (y \in \mathbf{R}_+),$$

by (4.19), we have

$$||h_2||_{p,\phi^{1-p}} = \left[\int_0^\infty \phi^{1-p}(y) h_2^p(y) dy \right]^{\frac{1}{p}} < M_1 ||f||_{p,\varphi} < \infty. \tag{4.36}$$

Definition 4.19 Define a Hardy-type integral operator of the first kind with the homogeneous kernel

$$T_1^{(2)} : L_{p,\varphi}(\mathbf{R}_+) \to L_{p,\phi^{1-p}}(\mathbf{R}_+)$$

as follows:
For any $f \in L_{p,\varphi}(\mathbf{R})$, there exists a unique representation

$$T_1^{(2)} f = h_2 \in L_{p,\phi^{1-p}}(\mathbf{R}_+),$$

satisfying

$$T_1^{(2)} f(y) = h_2(y),$$

for any $y \in \mathbf{R}_+$.

In view of (4.36), it follows that

$$\|T_1^{(2)} f\|_{p,\phi^{1-p}} = \|h_2\|_{p,\phi^{1-p}} \leq M_1 \|f\|_{p,\varphi},$$

and then the operator $T_1^{(2)}$ is bounded satisfying

$$\|T_1^{(2)}\| = \sup_{f(\neq\theta)\in L_{p,\varphi}(\mathbf{R}_+)} \frac{\|T_1^{(2)} f\|_{p,\phi^{1-p}}}{\|f\|_{p,\varphi}} \leq M_1.$$

If we define the formal inner product of $T_1^{(2)} f$ and g as follows:

$$(T_1^{(2)} f, g) = \int_0^\infty \left(\int_0^y k_\lambda(x, y) f(x) dx \right) g(y) dy,$$

we can then re-express Corollary 4.7 (for $\mu_1 = \mu$) using Remark 4.8 as follows:

Corollary 4.20 *Let M_1 be a constant. The following statements (i), (ii) and (iii) are equivalent:*

(i) For any $f(x) \geq 0$, $f \in L_{p,\varphi}(\mathbf{R}_+)$, $\|f\|_{p,\varphi} > 0$, we have the following inequality:

$$\|T_1^{(2)} f\|_{p,\phi^{1-p}} < M_1 \|f\|_{p,\varphi}. \tag{4.37}$$

(ii) For any $f(x), g(y) \geq 0$, $f \in L_{p,\varphi}(\mathbf{R}_+)$, $g \in L_{q,\phi}(\mathbf{R}_+)$, $\|f\|_{p,\varphi}, \|g\|_{q,\phi} > 0$, we have the following inequality:

$$(T_1^{(2)} f, g) < M_1 \|f\|_{p,\varphi} \|g\|_{q,\phi}. \tag{4.38}$$

(iii) $k_\lambda^{(1)}(\sigma) \leq M_1 (< \infty)$.
If statement (iii) holds true, then we have

$$\|T_1^{(2)}\| = k_\lambda^{(1)}(\sigma) \leq M_1.$$

Note. If $k_\lambda^{(1)}(\sigma) = 0$, then $T_1^{(2)} = \theta$ and $\|T_1^{(2)}\| = 0 = k_\lambda^{(1)}(\sigma) \leq M_1$.

Remark 4.21 If $h(u) = k_\lambda(u, 1)$, then Theorem 4.18 and Corollary 4.20 are equivalent.

(c) In view of Theorem 4.10 ($\sigma_1 = \sigma$), for $f \in L_{p,\varphi}(\mathbf{R}_+)$, setting

$$H_1(y) := \int_{\frac{1}{y}}^\infty h(xy) f(x) dx \ (y \in \mathbf{R}_+),$$

by (4.23), we obtain that

$$||H_1||_{p,\psi^{1-p}\circ f} = \left[\int_0^\infty \psi^{1-p}(y) H_1^p(y) dy \right]^{\frac{1}{p}} < M_2 ||f||_{p,\varphi} < \infty. \qquad (4.39)$$

Definition 4.22 Define a Hardy-type integral operator of the second kind with the nonhomogeneous kernel

$$T_2^{(1)} : L_{p,\varphi}(\mathbf{R}_+) \to L_{p,\psi^{1-p}}(\mathbf{R}_+)$$

as follows:
For any $f \in L_{p,\varphi}(\mathbf{R}_+)$, there exists a unique representation

$$T_2^{(1)} f = H_1 \in L_{p,\psi^{1-p}}(\mathbf{R}_+),$$

satisfying

$$T_2^{(1)} f(y) = H_1(y)$$

for any $y \in \mathbf{R}_+$.

In view of (4.39), it follows that

$$||T_2^{(1)} f||_{p,\psi^{1-p}} = ||H_1||_{p,\psi^{1-p}} \leq M_2 ||f||_{p,\varphi},$$

and then the operator $T_2^{(1)}$ is bounded satisfying

$$||T_2^{(1)}|| = \sup_{f(\neq\theta)\in L_{p,\varphi}(\mathbf{R}_+)} \frac{||T_2^{(1)} f||_{p,\psi^{1-p}}}{||f||_{p,\varphi}} \leq M_2.$$

If we define the formal inner product of $T_2^{(1)} f$ and g as follows:

$$(T_2^{(1)} f, g) = \int_0^\infty \left(\int_{\frac{1}{y}}^\infty h(xy) f(x) dx \right) g(y) dy,$$

we can then re-express Theorem 4.10 (for $\sigma_1 = \sigma$), using Remark 4.11, as follows:

Theorem 4.23 *Let M_2 be a constant. The following statements (i), (ii) and (iii) are equivalent:*
(i) For any $f(x) \geq 0, f \in L_{p,\varphi}(\mathbf{R}_+), ||f||_{p,\varphi} > 0$, we have the following inequality:

$$||T_2^{(1)} f||_{p,\psi^{1-p}} < M_2 ||f||_{p,\varphi}. \qquad (4.40)$$

(ii) For any $f(x), g(y) \geq 0, f \in L_{p,\varphi}(\mathbf{R}_+), g \in L_{q,\psi}(\mathbf{R}_+), ||f||_{p,\varphi}, ||g||_{q,\psi} > 0$, we have the following inequality:

$$(T_2^{(1)} f, g) < M_2 \|f\|_{p,\varphi} \|g\|_{q,\psi}. \tag{4.41}$$

(iii) $k_2(\sigma) \le M_2 \ (< \infty)$.
If statement (iii) holds true, then we have

$$\|T_2^{(1)}\| = k_2(\sigma) \le M_2.$$

Note. If $k_2(\sigma) = 0$, then $T_2^{(1)} = \theta$ and $\|T_2^{(1)}\| = 0 = k_2(\sigma) \le M_2$.
(d) In view of Corollary 4.14 ($\mu_1 = \mu$), for $f \in L_{p,\varphi}(\mathbf{R}_+)$, setting

$$H_2(y) := \int_y^\infty k_\lambda(x, y) f(x) dx \ (y \in \mathbf{R}_+),$$

by (4.29), we have

$$\|H_2\|_{p,\phi^{1-p}} = \left[\int_0^\infty \phi^{1-p}(y) H_2^p(y) dy\right]^{\frac{1}{p}} < M_2 \|f\|_{p,\varphi} < \infty. \tag{4.42}$$

Definition 4.24 Define a Hardy-type integral operator of the second kind with the homogeneous kernel
$$T_2^{(2)} : L_{p,\varphi}(\mathbf{R}_+) \to L_{p,\phi^{1-p}}(\mathbf{R}_+)$$

as follows:
For any $f \in L_{p,\varphi}(\mathbf{R})$, there exists a unique representation

$$T_2^{(2)} f = H_2 \in L_{p,\phi^{1-p}}(\mathbf{R}_+),$$

satisfying
$$T_2^{(2)} f(y) = H_2(y),$$

for any $y \in \mathbf{R}_+$.

In view of (4.42), it follows that

$$\|T_2^{(2)} f\|_{p,\phi^{1-p}} = \|H_2\|_{p,\phi^{1-p}} \le M_2 \|f\|_{p,\varphi},$$

and then the operator $T_2^{(2)}$ is bounded satisfying

$$\|T_2^{(2)}\| = \sup_{f(\ne \theta) \in L_{p,\varphi}(\mathbf{R}_+)} \frac{\|T_2^{(2)} f\|_{p,\phi^{1-p}}}{\|f\|_{p,\varphi}} \le M_2.$$

If we define the formal inner product of $T_1^{(2)} f$ and g as follows:

$$(T_2^{(2)} f, g) = \int_0^\infty \left(\int_y^\infty k_\lambda(x, y) f(x) dx \right) g(y) dy,$$

we can then re-express Corollary 4.14 (for $\mu_1 = \mu$) using Remark 4.15(a) as follows:

Corollary 4.25 *Let M_2 be a constant. The following statements (i), (ii) and (iii) are equivalent:*

(i) For any $f(x) \geq 0$, $f \in L_{p,\varphi}(\mathbf{R}_+)$, $||f||_{p,\varphi} > 0$, we have the following inequality:

$$||T_2^{(2)} f||_{p,\phi^{1-p}} < M_2 ||f||_{p,\varphi}. \tag{4.43}$$

(ii) For any $f(x)$, $g(y) \geq 0$, $f \in L_{p,\varphi}(\mathbf{R}_+)$, $g \in L_{q,\phi}(\mathbf{R}_+)$, $||f||_{p,\varphi}$, $||g||_{q,\phi} > 0$, we have the following inequality:

$$(T_2^{(2)} f, g) < M_2 ||f||_{p,\varphi} ||g||_{q,\phi}. \tag{4.44}$$

(iii) $k_\lambda^{(2)}(\sigma) \leq M_2(< \infty)$.
If statement (iii) holds true, then we have

$$||T_2^{(2)}|| = k_\lambda^{(2)}(\sigma) \leq M_2.$$

Note. If $k_\lambda^{(2)}(\sigma) = 0$, then $T_2^{(2)} = \theta$ and $||T_2^{(2)}|| = 0 = k_\lambda^{(2)}(\sigma) \leq M_2$.

Remark 4.26 If $h(u) = k_\lambda(u, 1)$, then Theorem 4.23 and Corollary 4.25 are equivalent.

Example 4.27 For $\beta > 0$, $\lambda > -\alpha$, setting

$$h(u) = k_\lambda(u, 1) = \frac{|\ln u|^\beta (\min\{u, 1\})^{\alpha+\gamma}}{|u^{\lambda+\alpha} - 1|(\max\{u, 1\})^\gamma} \ (u > 0),$$

we get

$$h(xy) = \frac{|\ln xy|^\beta (\min\{xy, 1\})^{\alpha+\gamma}}{|(xy)^{\lambda+\alpha} - 1|(\max\{xy, 1\})^\gamma},$$

$$k_\lambda(x, y) = \frac{|\ln \frac{x}{y}|^\beta (\min\{x, y\})^{\alpha+\gamma}}{|x^{\lambda+\alpha} - y^{\lambda+\alpha}|(\max\{x, y\})^\gamma},$$

and for $\sigma > -\alpha - \gamma$, we obtain

$$k_1(\sigma) = k_\lambda^{(1)}(\sigma) = \int_0^1 \frac{(-\ln u)^\beta u^{\alpha+\gamma}}{1 - u^{\lambda+\alpha}} u^{\sigma-1} du$$

$$= \int_0^1 (-\ln u)^\beta \sum_{k=0}^\infty u^{k(\lambda+\alpha)+\alpha+\gamma+\sigma-1} du$$

$$= \sum_{k=0}^{\infty} \int_0^1 (-\ln u)^{\beta} u^{k(\lambda+\alpha)+\alpha+\gamma+\sigma-1} du.$$

Setting

$$v = [k(\lambda + \alpha) + \alpha + \gamma + \sigma](-\ln u)$$

in the previous integral, we get

$$k_1(\sigma) = k_{\lambda}^{(1)}(\sigma) = \sum_{k=0}^{\infty} \frac{1}{[k(\lambda+\alpha)+\alpha+\gamma+\sigma]^{\beta+1}} \int_0^{\infty} v^{\beta} e^{-v} dv$$

$$= \frac{1}{(\lambda+\alpha)^{\beta+1}} \sum_{k=0}^{\infty} \frac{1}{(k+\frac{\alpha+\gamma+\sigma}{\lambda+\alpha})^{\beta+1}} \int_0^{\infty} v^{\beta} e^{-v} dv$$

$$= \frac{\Gamma(\beta+1)}{(\lambda+\alpha)^{\beta+1}} \zeta(\beta+1, \frac{\alpha+\gamma+\sigma}{\lambda+\alpha}) \in \mathbf{R}_+.$$

Then we have

$$||T_1^{(1)}|| = ||T_1^{(2)}|| = \frac{\Gamma(\beta+1)}{(\lambda+\alpha)^{\beta+1}} \zeta(\beta+1, \frac{\alpha+\gamma+\sigma}{\lambda+\alpha}). \tag{4.45}$$

For $\mu > -\alpha - \gamma$, setting $v = \frac{1}{u}$, it follows that

$$k_2(\sigma) = k_{\lambda}^{(2)}(\sigma) = \int_1^{\infty} \frac{(\ln u)^{\beta}}{(u^{\lambda+\alpha}-1)u^{\gamma}} u^{\sigma-1} du$$

$$= \int_0^1 \frac{(-\ln v)^{\beta}}{1-v^{\lambda+\alpha}} v^{\alpha+\gamma+\mu-1} dv$$

$$= \frac{\Gamma(\beta+1)}{(\lambda+\alpha)^{\beta+1}} \zeta(\beta+1, \frac{\alpha+\gamma+\mu}{\lambda+\alpha}) \in \mathbf{R}_+.$$

Then we obtain

$$||T_2^{(1)}|| = ||T_2^{(2)}|| = \frac{\Gamma(\beta+1)}{(\lambda+\alpha)^{\beta+1}} \zeta(\beta+1, \frac{\alpha+\gamma+\mu}{\lambda+\alpha}). \tag{4.46}$$

Note. If $\beta > 0, \lambda > -\alpha, \sigma, \mu > -\alpha - \gamma$, then $k_i(\sigma) = k_{\lambda}^{(i)}(\sigma) \in \mathbf{R}_+$ $(i = 1, 2)$.

Example 4.28 For $\beta > 0$, setting

$$h(u) = k_{\lambda}(u, 1) = \frac{|\ln u|^{\beta} \min\{u, 1\}}{|u-1|(\max\{u, 1\})^{\lambda}} \quad (u > 0),$$

we find

$$h(xy) = \frac{|\ln xy|^{\beta} \min\{xy, 1\}}{|xy - 1|(\max\{xy, 1\})^{\lambda}},$$

$$k_{\lambda}(x, y) = \frac{|\ln \frac{x}{y}|^{\beta} \min\{x, y\}}{|x - y|(\max\{x, y\})^{\lambda}},$$

and for $\sigma > -1$, by the Lebesgue term by term theorem, we obtain

$$k_1(\sigma) = k_{\lambda}^{(1)}(\sigma) = \int_0^1 h(u)u^{\sigma-1}du$$

$$= \int_0^1 \frac{(-\ln u)^{\beta}}{1-u}u^{\sigma}du = \int_0^1 (-\ln u)^{\beta} \sum_{k=0}^{\infty} u^{k+\sigma}du$$

$$= \sum_{k=0}^{\infty} \int_0^1 (-\ln u)^{\beta}u^{k+\sigma}du.$$

Setting

$$v = (k + \sigma + 1)(-\ln u)$$

in the previous integral, we obtain that

$$k_1(\sigma) = k_{\lambda}^{(1)}(\sigma) = \sum_{k=0}^{\infty} \frac{1}{(k + \sigma + 1)^{\beta+1}} \int_0^{\infty} v^{\beta}e^{-v}dv$$

$$= \Gamma(\beta + 1)\zeta(\beta + 1, \sigma + 1) \in \mathbf{R}_+.$$

Then we have

$$||T_1^{(1)}|| = ||T_1^{(2)}|| = \Gamma(\beta + 1)\zeta(\beta + 1, \sigma + 1). \tag{4.47}$$

For $\mu > -1$, setting $v = \frac{1}{u}$, it follows that

$$k_2(\sigma) = k_{\lambda}^{(2)}(\sigma) = \int_1^{\infty} \frac{(\ln u)^{\beta}}{(u - 1)u^{\lambda}}u^{\sigma-1}du$$

$$= \int_0^1 \frac{(-\ln v)^{\beta}}{1-v}v^{\mu}dv$$

$$= \Gamma(\beta + 1)\zeta(\beta + 1, \mu + 1) \in \mathbf{R}_+.$$

Then we get

$$||T_2^{(1)}|| = ||T_2^{(2)}|| = \Gamma(\beta + 1)\zeta(\beta + 1, \mu + 1). \tag{4.48}$$

Note. If $\beta > 0, \sigma, \mu > -1$, then $\lambda = \sigma + \mu > -2$, and $k_i(\sigma) = k_{\lambda}^{(i)}(\sigma) \in \mathbf{R}_+$ ($i = 1, 2$).

4.5 Hardy-Type Integral Inequalities with the Exponent Function as Interval Variables

For $a \in \mathbf{R}\backslash\{0\}, b > 0$, replacing x (resp. y) by e^{ax} (resp. e^{by}), then replacing $f(e^{ax})e^{ax}$ (resp. $g(e^{by})e^{by}$) by $f(x)$ (resp. $g(y)$), and $\frac{M}{|a|^{1/q}b^{1/p}}$ by M in Theorem 4.3, Corollary 4.7, Theorem 4.10 and Corollary 4.14, and by carrying out the corresponding calculations, we obtain the following theorem:

Theorem 4.29 *Let M_1 be a constant, and $a \neq 0, b > 0$. If $k_1(\sigma) > 0$, then the following statements (i), (ii) and (iii) are equivalent:*
   (i) For any $f(x) \geq 0$, satisfying

$$0 < \int_{-\infty}^{\infty} \left(\frac{f(x)}{e^{\sigma ax}}\right)^p dx < \infty,$$

we have the following inequality:

$$\left[\int_{-\infty}^{\infty} e^{p\sigma_1 by} \left(\int_{-\infty}^{e^{-by}} h(e^{ax+by}) f(x) dx\right)^p dy\right]^{\frac{1}{p}}$$

$$< M_1 \left[\int_{-\infty}^{\infty} \left(\frac{f(x)}{e^{\sigma ax}}\right)^p dx\right]^{\frac{1}{p}}. \tag{4.49}$$

   (ii) For any $f(x) \geq 0$, satisfying

$$0 < \int_{-\infty}^{\infty} \left(\frac{f(x)}{e^{\sigma ax}}\right)^p dx < \infty,$$

and $g(y) \geq 0$, satisfying

$$0 < \int_{-\infty}^{\infty} \left(\frac{g(y)}{e^{\sigma_1 by}}\right)^q dy < \infty,$$

we have the following inequality:

$$\int_{-\infty}^{\infty} g(y) \left(\int_{-\infty}^{e^{-by}} h(e^{ax+by}) f(x) dx\right) dy$$

$$< M_1 \left[\int_{-\infty}^{\infty} \left(\frac{f(x)}{e^{\sigma ax}}\right)^p dx\right]^{\frac{1}{p}} \left[\int_{-\infty}^{\infty} \left(\frac{g(y)}{e^{\sigma_1 by}}\right)^q dy\right]^{\frac{1}{q}}. \tag{4.50}$$

   (iii)

$$\sigma_1 = \sigma \text{ and } \frac{k_1(\sigma)}{|a|^{1/q}b^{1/p}} \leq M_1(< \infty).$$

If statement (iii) holds true, then the constant $M_1 = \frac{k_1(\sigma)}{|a|^{1/q}b^{1/p}}$ ($\in \mathbf{R}_+$) *in (4.49) and*
(4.50) is the best possible.

In particular, for $\sigma_1 = \sigma$,

$$h(u) = \frac{|\ln u|^\beta (\min\{u, 1\})^{\alpha+\gamma}}{|u^{\lambda+\alpha} - 1|(\max\{u, 1\})^\gamma}$$

($\beta > 0, \lambda > -\alpha, \sigma > -\alpha - \gamma$), by Example 4.27, we have the following equivalent
inequalities with

$$\frac{\Gamma(\beta+1)}{|a|^{1/q}b^{1/p}(\lambda+\alpha)^{\beta+1}}\zeta\left(\beta+1, \frac{\alpha+\gamma+\sigma}{\lambda+\alpha}\right)$$

being the best possible constant factor:

$$\left\{\int_{-\infty}^\infty e^{p\sigma by}\left[\int_{-\infty}^{e^{-by}} \frac{|ax + by|^\beta (\min\{e^{ax+by}, 1\})^{\alpha+\gamma} f(x)}{|e^{(ax+by)(\lambda+\alpha)} - 1|(\max\{e^{ax+by}, 1\})^\gamma}dx\right]^p dy\right\}^{\frac{1}{p}}$$

$$< \frac{\Gamma(\beta+1)}{|a|^{1/q}b^{1/p}(\lambda+\alpha)^{\beta+1}}\zeta\left(\beta+1, \frac{\alpha+\gamma+\sigma}{\lambda+\alpha}\right)\left[\int_{-\infty}^\infty \left(\frac{f(x)}{e^{\sigma ax}}\right)^p dx\right]^{\frac{1}{p}}, \quad (4.51)$$

$$\int_{-\infty}^\infty g(y)\left[\int_{-\infty}^{e^{-by}} \frac{|ax + by|^\beta (\min\{e^{ax+by}, 1\})^{\alpha+\gamma}}{|e^{(ax+by)(\lambda+\alpha)} - 1|(\max\{e^{ax+by}, 1\})^\gamma}f(x)dx\right] dy$$

$$< \frac{\Gamma(\beta+1)}{|a|^{1/q}b^{1/p}(\lambda+\alpha)^{\beta+1}}\zeta\left(\beta+1, \frac{\alpha+\gamma+\sigma}{\lambda+\alpha}\right)$$

$$\times \left[\int_{-\infty}^\infty \left(\frac{f(x)}{e^{\sigma ax}}\right)^p dx\right]^{\frac{1}{p}}\left[\int_{-\infty}^\infty \left(\frac{g(y)}{e^{\sigma by}}\right)^q dy\right]^{\frac{1}{q}}. \quad (4.52)$$

Remark 4.30 In Theorem 4.29, if $\sigma_1 = \sigma, b < 0$, then replacing $-b$ by b (> 0),
we have the following equivalent inequalities with the best possible constant factor
$\frac{k_1(\sigma)}{|a|^{1/q}b^{1/p}}$:

$$\left[\int_{-\infty}^\infty e^{-p\sigma by}\left(\int_{e^{by}}^\infty h(e^{ax-by})f(x)dx\right)^p dy\right]^{\frac{1}{p}}$$

$$< \frac{k_1(\sigma)}{|a|^{1/q}b^{1/p}}\left[\int_{-\infty}^\infty \left(\frac{f(x)}{e^{\sigma ax}}\right)^p dx\right]^{\frac{1}{p}}. \quad (4.53)$$

$$\int_{-\infty}^\infty g(y)\left(\int_{e^{by}}^\infty h(e^{ax-by})f(x)dx\right) dy$$

$$< \frac{k_1(\sigma)}{|a|^{1/q}b^{1/p}}\left[\int_{-\infty}^\infty \left(\frac{f(x)}{e^{\sigma ax}}\right)^p dx\right]^{\frac{1}{p}}\left[\int_{-\infty}^\infty \left(\frac{g(y)}{e^{-\sigma by}}\right)^q dy\right]^{\frac{1}{q}}. \quad (4.54)$$

In particular, for

$$h(u) = \frac{|\ln u|^\beta \min\{u, 1\}}{|u - 1|(\max\{u, 1\})^\lambda}$$

($\beta > 0, \sigma > -1$), by Example 4.28, we have the following equivalent inequalities with

$$\frac{\Gamma(\beta + 1)}{|a|^{1/q} b^{1/p}} \zeta(\beta + 1, \sigma + 1)$$

being the best possible constant factor:

$$\left\{ \int_{-\infty}^{\infty} e^{-p\sigma by} \left[\int_{e^{by}}^{\infty} \frac{|ax - by|^\beta \min\{e^{ax-by}, 1\} f(x)}{|e^{ax-by} - 1|(\max\{e^{ax-by}, 1\})^\lambda} dx \right]^p dy \right\}^{\frac{1}{p}}$$

$$< \frac{\Gamma(\beta + 1)}{|a|^{1/q} b^{1/p}} \zeta(\beta + 1, \sigma + 1) \left[\int_{-\infty}^{\infty} \left(\frac{f(x)}{e^{\sigma ax}} \right)^p dx \right]^{\frac{1}{p}}, \tag{4.55}$$

$$\int_{-\infty}^{\infty} g(y) \left[\int_{e^{by}}^{\infty} \frac{|ax - by|^\beta \min\{e^{ax-by}, 1\}}{|e^{ax-by} - 1|(\max\{e^{ax-by}, 1\})^\lambda} f(x) dx \right] dy$$

$$< \frac{\Gamma(\beta + 1)}{|a|^{1/q} b^{1/p}} \zeta(\beta + 1, \sigma + 1)$$

$$\times \left[\int_{-\infty}^{\infty} \left(\frac{f(x)}{e^{\sigma ax}} \right)^p dx \right]^{\frac{1}{p}} \left[\int_{-\infty}^{\infty} \left(\frac{g(y)}{e^{-\sigma by}} \right)^q dy \right]^{\frac{1}{q}}. \tag{4.56}$$

Corollary 4.31 *Suppose that M_1 is a constant, and $a \neq 0, b > 0$. If*

$$k_\lambda^{(1)}(\sigma) = \int_0^1 k_\lambda(u, 1) u^{\sigma-1} du > 0,$$

then the following statements (i), (ii) and (iii) are equivalent:
(i) For any $f(x) \geq 0$, satisfying

$$0 < \int_{-\infty}^{\infty} \left(\frac{f(x)}{e^{\sigma ax}} \right)^p dx < \infty,$$

we have the following inequality:

$$\left[\int_{-\infty}^{\infty} e^{p\mu_1 by} \left(\int_{-\infty}^{e^{by}} k_\lambda(e^{ax}, e^{by}) f(x) dx \right)^p dy \right]^{\frac{1}{p}}$$

$$< M_1 \left[\int_{-\infty}^{\infty} \left(\frac{f(x)}{e^{\sigma ax}} \right)^p dx \right]^{\frac{1}{p}}. \tag{4.57}$$

(ii) For any $f(x) \geq 0$, satisfying

$$0 < \int_{-\infty}^{\infty} \left(\frac{f(x)}{e^{\sigma ax}} \right)^p dx < \infty,$$

and $g(y) \geq 0$, satisfying

$$0 < \int_{-\infty}^{\infty} \left(\frac{g(y)}{e^{\mu_1 by}} \right)^q dy < \infty,$$

we have the following inequality:

$$\int_{-\infty}^{\infty} g(y) \left(\int_{-\infty}^{e^{by}} k_\lambda(e^{ax}, e^{by}) f(x) dx \right) dy$$

$$< M_1 \left[\int_{-\infty}^{\infty} \left(\frac{f(x)}{e^{\sigma ax}} \right)^p dx \right]^{\frac{1}{p}} \left[\int_{-\infty}^{\infty} \left(\frac{g(y)}{e^{\mu_1 by}} \right)^q dy \right]^{\frac{1}{q}}. \qquad (4.58)$$

(iii)

$$\mu_1 = \mu \ \text{and} \ \frac{k_\lambda^{(1)}(\sigma)}{|a|^{1/q} b^{1/p}} \leq M_1 \ (< \infty).$$

If statement (iii) holds true, then the constant $M_1 = \frac{k_\lambda^{(1)}(\sigma)}{|a|^{1/q} b^{1/p}} \ (\in \mathbf{R}_+)$ in (4.57) and (4.58) is the best possible.

In particular, for $\mu_1 = \mu$,

$$h(u) = \frac{|\ln u|^\beta (\min\{u, 1\})^{\alpha + \gamma}}{|u^{\lambda + \alpha} - 1| (\max\{u, 1\})^\gamma}$$

$(\beta > 0, \lambda > -\alpha, \sigma > -\alpha - \gamma)$, by Example 4.27, we have the following equivalent inequalities with

$$\frac{\Gamma(\beta + 1)}{|a|^{1/q} b^{1/p} (\lambda + \alpha)^{\beta + 1}} \zeta \left(\beta + 1, \frac{\alpha + \gamma + \sigma}{\lambda + \alpha} \right)$$

being the best possible constant factor:

$$\left\{ \int_{-\infty}^{\infty} e^{p\mu by} \left[\int_{-\infty}^{e^{by}} \frac{|ax + by|^\beta (\min\{e^{ax}, e^{by}\})^{\alpha + \gamma} f(x)}{|e^{(\lambda + \alpha)ax} - e^{(\lambda + \alpha)by}| (\max\{e^{ax}, e^{by}\})^\gamma} dx \right]^p dy \right\}^{\frac{1}{p}}$$

$$< \frac{\Gamma(\beta + 1)}{|a|^{1/q} b^{1/p} (\lambda + \alpha)^{\beta + 1}} \zeta \left(\beta + 1, \frac{\alpha + \gamma + \sigma}{\lambda + \alpha} \right) \left[\int_{-\infty}^{\infty} \left(\frac{f(x)}{e^{\sigma ax}} \right)^p dx \right]^{\frac{1}{p}}, \quad (4.59)$$

$$\int_{-\infty}^{\infty} g(y) \left[\int_{-\infty}^{e^{by}} \frac{|ax + by|^{\beta} (\min\{e^{ax}, e^{by}\})^{\alpha + \gamma}}{|e^{(\lambda + \alpha)ax} - e^{(\lambda + \alpha)by}| (\max\{e^{ax}, e^{by}\})^{\gamma}} f(x)dx \right] dy$$

$$< \frac{\Gamma(\beta + 1)}{|a|^{1/q} b^{1/p} (\lambda + \alpha)^{\beta + 1}} \zeta \left(\beta + 1, \frac{\alpha + \gamma + \sigma}{\lambda + \alpha} \right)$$

$$\times \left[\int_{-\infty}^{\infty} \left(\frac{f(x)}{e^{\sigma ax}} \right)^{p} dx \right]^{\frac{1}{p}} \left[\int_{-\infty}^{\infty} \left(\frac{g(y)}{e^{\mu by}} \right)^{q} dy \right]^{\frac{1}{q}}. \tag{4.60}$$

Remark 4.32 In Corollary 4.31, if $\mu_1 = \mu$, $b < 0$, then replacing $-b$ by b, we have the following equivalent inequalities with the best possible constant factor $\frac{k_{\lambda}^{(1)}(\sigma)}{|a|^{1/q} b^{1/p}}$:

$$\left[\int_{-\infty}^{\infty} e^{-p\mu by} \left(\int_{e^{-by}}^{\infty} k_{\lambda}(e^{ax}, e^{-by}) f(x)dx \right)^{p} dy \right]^{\frac{1}{p}}$$

$$< \frac{k_{\lambda}^{(1)}(\sigma)}{|a|^{1/q} b^{1/p}} \left[\int_{-\infty}^{\infty} \left(\frac{f(x)}{e^{\sigma ax}} \right)^{p} dx \right]^{\frac{1}{p}}. \tag{4.61}$$

$$\int_{-\infty}^{\infty} g(y) \left(\int_{e^{-by}}^{\infty} k_{\lambda}(e^{ax}, e^{-by}) f(x)dx \right) dy$$

$$< \frac{k_{\lambda}^{(1)}(\sigma)}{|a|^{1/q} b^{1/p}} \left[\int_{-\infty}^{\infty} \left(\frac{f(x)}{e^{\sigma ax}} \right)^{p} dx \right]^{\frac{1}{p}} \left[\int_{-\infty}^{\infty} \left(\frac{g(y)}{e^{-\mu by}} \right)^{q} dy \right]^{\frac{1}{q}}. \tag{4.62}$$

In particular, for

$$k_{\lambda}(u, 1) = \frac{|\ln u|^{\beta} \min\{u, 1\}}{|u - 1| (\max\{u, 1\})^{\lambda}}$$

$(\beta > 0, \sigma > -1)$, by Example 4.28, we have the following equivalent inequalities with

$$\frac{\Gamma(\beta + 1)}{|a|^{1/q} b^{1/p}} \zeta(\beta + 1, \sigma + 1)$$

being the best possible constant factor:

$$\left\{ \int_{-\infty}^{\infty} e^{-p\mu by} \left[\int_{e^{-by}}^{\infty} \frac{|ax - by|^{\beta} \min\{e^{ax}, e^{-by}\} f(x)}{|e^{ax} - e^{-by}| (\max\{e^{ax}, e^{-by}\})^{\lambda}} dx \right]^{p} dy \right\}^{\frac{1}{p}}$$

$$< \frac{\Gamma(\beta + 1)}{|a|^{1/q} b^{1/p}} \zeta(\beta + 1, \sigma + 1) \left[\int_{-\infty}^{\infty} \left(\frac{f(x)}{e^{\sigma ax}} \right)^{p} dx \right]^{\frac{1}{p}}, \tag{4.63}$$

$$\int_{-\infty}^{\infty} g(y) \left[\int_{e^{-by}}^{\infty} \frac{|ax - by|^{\beta} \min\{e^{ax}, e^{-by}\}}{|e^{ax} - e^{-by}| (\max\{e^{ax}, e^{-by}\})^{\lambda}} f(x)dx \right] dy$$

$$< \frac{\Gamma(\beta+1)}{|a|^{1/q}b^{1/p}} \zeta(\beta+1,\sigma+1)$$

$$\times \left[\int_{-\infty}^{\infty} \left(\frac{f(x)}{e^{\sigma a x}}\right)^p dx\right]^{\frac{1}{p}} \left[\int_{-\infty}^{\infty} \left(\frac{g(y)}{e^{-\mu b y}}\right)^q dy\right]^{\frac{1}{q}}. \tag{4.64}$$

Theorem 4.33 *Let M_2 be a constant, and $a \neq 0, b > 0$. If $k_2(\sigma) > 0$, then the following statements (i), (ii) and (iii) are equivalent:*

(i) For any $f(x) \geq 0$, satisfying

$$0 < \int_{-\infty}^{\infty} \left(\frac{f(x)}{e^{\sigma a x}}\right)^p dx < \infty,$$

we have the following inequality:

$$\left[\int_{-\infty}^{\infty} e^{p\sigma_1 by} \left(\int_{e^{-by}}^{\infty} h(e^{ax+by})f(x)dx\right)^p dy\right]^{\frac{1}{p}}$$

$$< M_2 \left[\int_{-\infty}^{\infty} \left(\frac{f(x)}{e^{\sigma a x}}\right)^p dx\right]^{\frac{1}{p}}. \tag{4.65}$$

(ii) For any $f(x) \geq 0$, satisfying

$$0 < \int_{-\infty}^{\infty} \left(\frac{f(x)}{e^{\sigma a x}}\right)^p dx < \infty,$$

and $g(y) \geq 0$, satisfying

$$0 < \int_{-\infty}^{\infty} \left(\frac{g(y)}{e^{\sigma_1 by}}\right)^q dy < \infty,$$

we have the following inequality:

$$\int_{-\infty}^{\infty} g(y) \left(\int_{e^{-by}}^{\infty} h(e^{ax+by})f(x)dx\right) dy$$

$$< M_2 \left[\int_{-\infty}^{\infty} \left(\frac{f(x)}{e^{\sigma a x}}\right)^p dx\right]^{\frac{1}{p}} \left[\int_{-\infty}^{\infty} \left(\frac{g(y)}{e^{\sigma_1 by}}\right)^q dy\right]^{\frac{1}{q}}. \tag{4.66}$$

(iii) $\sigma_1 = \sigma$, and $\frac{k_2(\sigma)}{|a|^{1/q}b^{1/p}} \leq M_2 \ (< \infty)$.

If statement (iii) holds true, then the constant $M_2 = \frac{k_2(\sigma)}{|a|^{1/q}b^{1/p}} \ (\in \mathbf{R}_+)$ in (4.65) and (4.66) is the best possible.

In particular, for $\sigma_1 = \sigma$,

$$h(u) = \frac{|\ln u|^\beta (\min\{u, 1\})^{\alpha+\gamma}}{|u^{\lambda+\alpha} - 1|(\max\{u, 1\})^\gamma}$$

$(\beta > 0, \lambda > -\alpha, \mu > -\alpha - \gamma)$, by Example 4.27, we have the following equivalent inequalities with

$$\frac{\Gamma(\beta+1)}{|a|^{1/q}b^{1/p}(\lambda+\alpha)^{\beta+1}}\zeta\left(\beta+1, \frac{\alpha+\gamma+\mu}{\lambda+\alpha}\right)$$

being the best possible constant factor:

$$\left\{\int_{-\infty}^\infty e^{p\sigma by}\left[\int_{e^{-by}}^\infty \frac{|ax+by|^\beta(\min\{e^{ax+by}, 1\})^{\alpha+\gamma}f(x)}{|e^{(ax+by)(\lambda+\alpha)} - 1|(\max\{e^{ax+by}, 1\})^\gamma}dx\right]^p dy\right\}^{\frac{1}{p}}$$

$$< \frac{\Gamma(\beta+1)}{|a|^{1/q}b^{1/p}(\lambda+\alpha)^{\beta+1}}\zeta\left(\beta+1, \frac{\alpha+\gamma+\mu}{\lambda+\alpha}\right)\left[\int_{-\infty}^\infty \left(\frac{f(x)}{e^{\sigma ax}}\right)^p dx\right]^{\frac{1}{p}}, \quad (4.67)$$

$$\int_{-\infty}^\infty g(y)\left[\int_{e^{-by}}^\infty \frac{|ax+by|^\beta(\min\{e^{ax+by}, 1\})^{\alpha+\gamma}}{|e^{(ax+by)(\lambda+\alpha)} - 1|(\max\{e^{ax+by}, 1\})^\gamma}f(x)dx\right]dy$$

$$< \frac{\Gamma(\beta+1)}{|a|^{1/q}b^{1/p}(\lambda+\alpha)^{\beta+1}}\zeta\left(\beta+1, \frac{\alpha+\gamma+\mu}{\lambda+\alpha}\right)$$

$$\times \left[\int_{-\infty}^\infty \left(\frac{f(x)}{e^{\sigma ax}}\right)^p dx\right]^{\frac{1}{p}}\left[\int_{-\infty}^\infty \left(\frac{g(y)}{e^{\sigma by}}\right)^q dy\right]^{\frac{1}{q}}. \quad (4.68)$$

Remark 4.34 In Theorem 4.33, if $\sigma_1 = \sigma, b < 0$, then replacing $-b$ by $b(> 0)$, we have the following equivalent inequalities with the best possible constant factor $\frac{k_2(\sigma)}{|a|^{1/q}b^{1/p}}$:

$$\left[\int_{-\infty}^\infty e^{-p\sigma by}\left(\int_{-\infty}^{e^{by}} h(e^{ax-by})f(x)dx\right)^p dy\right]^{\frac{1}{p}}$$

$$< \frac{k_2(\sigma)}{|a|^{1/q}b^{1/p}}\left[\int_{-\infty}^\infty \left(\frac{f(x)}{e^{\sigma ax}}\right)^p dx\right]^{\frac{1}{p}}. \quad (4.69)$$

$$\int_{-\infty}^\infty g(y)\left(\int_{-\infty}^{e^{by}} h(e^{ax-by})f(x)dx\right)dy$$

$$< \frac{k_2(\sigma)}{|a|^{1/q}b^{1/p}}\left[\int_{-\infty}^\infty \left(\frac{f(x)}{e^{\sigma ax}}\right)^p dx\right]^{\frac{1}{p}}\left[\int_{-\infty}^\infty \left(\frac{g(y)}{e^{-\sigma by}}\right)^q dy\right]^{\frac{1}{q}}. \quad (4.70)$$

In particular, for

$$h(u) = \frac{|\ln u|^\beta \min\{u, 1\}}{|u - 1|(\max\{u, 1\})^\lambda}$$

($\beta > 0, \mu > -1$), by Example 4.28, we have the following equivalent inequalities with

$$\frac{\Gamma(\beta + 1)}{|a|^{1/q}b^{1/p}}\zeta(\beta + 1, \mu + 1)$$

being the best possible constant factor:

$$\left\{\int_{-\infty}^{\infty} e^{-p\sigma by}\left[\int_{-\infty}^{e^{by}} \frac{|ax - by|^\beta \min\{e^{ax-by}, 1\}}{|e^{ax-by} - 1|(\max\{e^{ax-by}, 1\})^\lambda} f(x)dx\right]^p dy\right\}^{\frac{1}{p}}$$
$$< \frac{\Gamma(\beta + 1)}{|a|^{1/q}b^{1/p}}\zeta(\beta + 1, \mu + 1)\left[\int_{-\infty}^{\infty} \left(\frac{f(x)}{e^{\sigma ax}}\right)^p dx\right]^{\frac{1}{p}}, \tag{4.71}$$

$$\int_{-\infty}^{\infty} g(y)\left[\int_{-\infty}^{e^{by}} \frac{|ax - by|^\beta \min\{e^{ax-by}, 1\}}{|e^{ax-by} - 1|(\max\{e^{ax-by}, 1\})^\lambda} f(x)dx\right] dy$$
$$< \frac{\Gamma(\beta + 1)}{|a|^{1/q}b^{1/p}}\zeta(\beta + 1, \mu + 1)$$
$$\times \left[\int_{-\infty}^{\infty} \left(\frac{f(x)}{e^{\sigma ax}}\right)^p dx\right]^{\frac{1}{p}}\left[\int_{-\infty}^{\infty} \left(\frac{g(y)}{e^{-\sigma by}}\right)^q dy\right]^{\frac{1}{q}}. \tag{4.72}$$

Corollary 4.35 *Let M_2 be a constant, and $a \neq 0, b > 0$. If*

$$k_\lambda^{(2)}(\sigma) = \int_1^{\infty} k_\lambda(u, 1)u^{\sigma-1}du > 0,$$

then the following statements (i), (ii) and (iii) are equivalent:
  (i) For any $f(x) \geq 0$, satisfying

$$0 < \int_{-\infty}^{\infty} \left(\frac{f(x)}{e^{\sigma ax}}\right)^p dx < \infty,$$

we have the following inequality:

$$\left[\int_{-\infty}^{\infty} e^{p\mu_1 by}\left(\int_{e^{by}}^{\infty} k_\lambda(e^{ax}, e^{by})f(x)dx\right)^p dy\right]^{\frac{1}{p}}$$
$$< M_2\left[\int_{-\infty}^{\infty} \left(\frac{f(x)}{e^{\sigma ax}}\right)^p dx\right]^{\frac{1}{p}}. \tag{4.73}$$

(ii) For any $f(x) \geq 0$, satisfying

$$0 < \int_{-\infty}^{\infty} \left(\frac{f(x)}{e^{\sigma ax}} \right)^p dx < \infty,$$

and $g(y) \geq 0$, satisfying

$$0 < \int_{-\infty}^{\infty} \left(\frac{g(y)}{e^{\mu_1 by}} \right)^q dy < \infty,$$

we have the following inequality:

$$\int_{-\infty}^{\infty} g(y) \left(\int_{e^{by}}^{\infty} k_\lambda(e^{ax}, e^{by}) f(x) dx \right) dy$$

$$< M_2 \left[\int_{-\infty}^{\infty} \left(\frac{f(x)}{e^{\sigma ax}} \right)^p dx \right]^{\frac{1}{p}} \left[\int_{-\infty}^{\infty} \left(\frac{g(y)}{e^{\mu_1 by}} \right)^q dy \right]^{\frac{1}{q}}. \qquad (4.74)$$

(iii)

$$\mu_1 = \mu \;\; and \;\; \frac{k_\lambda^{(2)}(\sigma)}{|a|^{1/q} b^{1/p}} \leq M_2 \; (< \infty).$$

If statement (iii) holds true, then the constant $M_2 = \frac{k_\lambda^{(2)}(\sigma)}{|a|^{1/q} b^{1/p}}$ $(\in \mathbf{R}_+)$ in (4.73) and (4.74) is the best possible.

In particular, for $\mu_1 = \mu$,

$$h(u) = \frac{|\ln u|^\beta (\min\{u, 1\})^{\alpha + \gamma}}{|u^{\lambda + \alpha} - 1| (\max\{u, 1\})^\gamma}$$

$(\beta > 0, \lambda > -\alpha, \mu > -\alpha - \gamma)$, by Example 4.27, we have the following equivalent inequalities with

$$\frac{\Gamma(\beta + 1)}{|a|^{1/q} b^{1/p} (\lambda + \alpha)^{\beta + 1}} \zeta \left(\beta + 1, \frac{\alpha + \gamma + \mu}{\lambda + \alpha} \right)$$

being the best possible constant factor:

$$\left\{ \int_{-\infty}^{\infty} e^{p\sigma by} \left[\int_{e^{by}}^{\infty} \frac{|ax + by|^\beta (\min\{e^{ax}, e^{by}\})^{\alpha + \gamma} f(x)}{|e^{(\lambda + \alpha)ax} - e^{(\lambda + \alpha)by}| (\max\{e^{ax}, e^{by}\})^\gamma} dx \right]^p dy \right\}^{\frac{1}{p}}$$

$$< \frac{\Gamma(\beta + 1)}{|a|^{1/q} b^{1/p} (\lambda + \alpha)^{\beta + 1}} \zeta \left(\beta + 1, \frac{\alpha + \gamma + \mu}{\lambda + \alpha} \right) \left[\int_{-\infty}^{\infty} \left(\frac{f(x)}{e^{\sigma ax}} \right)^p dx \right]^{\frac{1}{p}}, \qquad (4.75)$$

$$\int_{-\infty}^{\infty} g(y) \left[\int_{e^{by}}^{\infty} \frac{|ax + by|^{\beta} (\min\{e^{ax}, e^{by}\})^{\alpha+\gamma}}{|e^{(\lambda+\alpha)ax} - e^{(\lambda+\alpha)by}|(\max\{e^{ax}, e^{by}\})^{\gamma}} f(x) dx \right] dy$$

$$< \frac{\Gamma(\beta+1)}{|a|^{1/q} b^{1/p} (\lambda+\alpha)^{\beta+1}} \zeta\left(\beta+1, \frac{\alpha+\gamma+\mu}{\lambda+\alpha}\right)$$

$$\times \left[\int_{-\infty}^{\infty} \left(\frac{f(x)}{e^{\sigma ax}}\right)^p dx \right]^{\frac{1}{p}} \left[\int_{-\infty}^{\infty} \left(\frac{g(y)}{e^{\mu by}}\right)^q dy \right]^{\frac{1}{q}}. \tag{4.76}$$

Remark 4.36 In Corollary 4.35, if $\mu_1 = \mu, b < 0$, then replacing $-b$ by $b(> 0)$, we have the following equivalent inequalities with the best possible constant factor $\frac{k_{\lambda}^{(2)}(\sigma)}{|a|^{1/q} b^{1/p}}$:

$$\left[\int_{-\infty}^{\infty} e^{-p\mu by} \left(\int_{-\infty}^{e^{-by}} k_{\lambda}(e^{ax}, e^{-by}) f(x) dx \right)^p dy \right]^{\frac{1}{p}}$$

$$< \frac{k_{\lambda}^{(2)}(\sigma)}{|a|^{1/q} b^{1/p}} \left[\int_{-\infty}^{\infty} \left(\frac{f(x)}{e^{\sigma ax}}\right)^p dx \right]^{\frac{1}{p}}. \tag{4.77}$$

$$\int_{-\infty}^{\infty} g(y) \left(\int_{-\infty}^{e^{-by}} k_{\lambda}(e^{ax}, e^{-by}) f(x) dx \right) dy$$

$$< \frac{k_{\lambda}^{(2)}(\sigma)}{|a|^{1/q} b^{1/p}} \left[\int_{-\infty}^{\infty} \left(\frac{f(x)}{e^{\sigma ax}}\right)^p dx \right]^{\frac{1}{p}} \left[\int_{-\infty}^{\infty} \left(\frac{g(y)}{e^{-\mu by}}\right)^q dy \right]^{\frac{1}{q}}. \tag{4.78}$$

In particular, for

$$k_{\lambda}(u, 1) = \frac{|\ln u|^{\beta} \min\{u, 1\}}{|u - 1|(\max\{u, 1\})^{\lambda}}$$

$(\beta > 0, \mu > -1)$, by Example 4.28, we have the following equivalent inequalities with

$$\frac{\Gamma(\beta+1)}{|a|^{1/q} b^{1/p}} \zeta(\beta+1, \mu+1)$$

being the best possible constant factor:

$$\left\{ \int_{-\infty}^{\infty} e^{-p\mu by} \left[\int_{-\infty}^{e^{-by}} \frac{|ax - by|^{\beta} \min\{e^{ax}, e^{-by}\} f(x)}{|e^{ax} - e^{-by}|(\max\{e^{ax}, e^{-by}\})^{\lambda}} dx \right]^p dy \right\}^{\frac{1}{p}}$$

$$< \frac{\Gamma(\beta+1)}{|a|^{1/q} b^{1/p}} \zeta(\beta+1, \mu+1) \left[\int_{-\infty}^{\infty} \left(\frac{f(x)}{e^{\sigma ax}}\right)^p dx \right]^{\frac{1}{p}}, \tag{4.79}$$

$$\int_{-\infty}^{\infty} g(y) \left[\int_{-\infty}^{e^{-by}} \frac{|ax - by|^{\beta} \min\{e^{ax}, e^{-by}\} f(x)}{|e^{ax} - e^{-by}|(\max\{e^{ax}, e^{-by}\})^{\lambda}} dx \right] dy$$

$$< \frac{\Gamma(\beta + 1)}{|a|^{1/q} b^{1/p}} \zeta(\beta + 1, \mu + 1)$$

$$\times \left[\int_{-\infty}^{\infty} \left(\frac{f(x)}{e^{\sigma ax}} \right)^{p} dx \right]^{\frac{1}{p}} \left[\int_{-\infty}^{\infty} \left(\frac{g(y)}{e^{-\mu by}} \right)^{q} dy \right]^{\frac{1}{q}}. \qquad (4.80)$$

References

1. Kuang, J.C.: Real and Functional Analysis (continuation) (sec. vol.). Higher Education Press, Beijing, China (2015)
2. Kuang, J.C.: Applied Inequalities. Shangdong Science and Technology Press, Jinan, China (2004)

Chapter 5
Equivalent Property of the Reverse Hardy-Type Integral Inequalities

In this chapter, we obtain a few equivalent statements of two kinds of reverse Hardy-type integral inequalities with a general nonhomogeneous kernel related to certain parameters. Two kinds of reverse Hardy-type integral inequalities with a general homogeneous kernel are deduced. We also consider a few particular examples involving the extended Hurwitz zeta function in the form of applications, as well as two kinds of reverse Hardy-type integral inequalities in the whole plane.

5.1 Two Lemmas

In the sequel, we assume that $0 < p < 1$ $(q < 0)$, $\frac{1}{p} + \frac{1}{q} = 1$, $\sigma_1, \sigma, \mu \in \mathbf{R}$, $\sigma + \mu = \lambda$, and $h(u)$ is a nonnegative measurable function in $(0, \infty)$, such that

$$k_1(\sigma) = \int_0^1 h(u)u^{\sigma-1}du \ (\geq 0), \tag{5.1}$$

$$k_2(\sigma) = \int_1^\infty h(u)u^{\sigma-1}du \ (\geq 0). \tag{5.2}$$

Lemma 5.1 *If $k_1(\sigma) < \infty$, and if there exists a constant $M_1 > 0$, such that for any nonnegative measurable functions $f(x)$ and $g(y)$ in $(0, \infty)$, the following inequality*

$$\int_0^\infty g(y) \left(\int_0^{\frac{1}{y}} h(xy)f(x)dx \right) dy$$

$$\geq M_1 \left[\int_0^\infty x^{p(1-\sigma)-1} f^p(x)dx \right]^{\frac{1}{p}} \left[\int_0^\infty y^{q(1-\sigma_1)-1} g^q(y)dy \right]^{\frac{1}{q}} \tag{5.3}$$

© The Author(s), under exclusive licence to Springer Nature Switzerland AG 2019
B. Yang and M. Th. Rassias, *On Hilbert-Type and Hardy-Type Integral Inequalities and Applications*, SpringerBriefs in Mathematics,
https://doi.org/10.1007/978-3-030-29268-3_5

holds true, then we have

$$\sigma_1 = \sigma \ \ and \ \ k_1(\sigma) \geq M_1 \ (> 0).$$

Proof If $\sigma_1 < \sigma$, then for $n \in \mathbf{N}$, we set the following functions:

$$f_n(x) = \begin{cases} x^{\sigma + \frac{1}{pn} - 1}, 0 < x \leq 1 \\ 0, x > 1 \end{cases},$$

$$g_n(y) = \begin{cases} 0, 0 < y < 1 \\ y^{\sigma_1 - \frac{1}{qn} - 1}, y \geq 1 \end{cases},$$

and find

$$J_1 = \left[\int_0^\infty x^{p(1-\sigma)-1} f_n^p(x) dx \right]^{\frac{1}{p}} \left[\int_0^\infty y^{q(1-\sigma_1)-1} g_n^q(y) dy \right]^{\frac{1}{q}}$$

$$= \left(\int_0^1 x^{\frac{1}{n}-1} dx \right)^{\frac{1}{p}} \left(\int_1^\infty y^{-\frac{1}{n}-1} dy \right)^{\frac{1}{q}} = n.$$

For fixed y, setting $u = xy$, we obtain

$$I_1 = \int_0^\infty g_n(y) \left(\int_0^{\frac{1}{y}} h(xy) f_n(x) dx \right) dy$$

$$= \int_1^\infty \left(\int_0^{\frac{1}{y}} h(xy) x^{\sigma + \frac{1}{pn} - 1} dx \right) y^{\sigma_1 - \frac{1}{qn} - 1} dy$$

$$= \int_1^\infty y^{(\sigma_1 - \sigma) - \frac{1}{n} - 1} dy \int_0^1 h(u) u^{\sigma + \frac{1}{pn} - 1} du$$

$$\leq \frac{1}{\sigma - \sigma_1 + \frac{1}{n}} \int_0^1 h(u) u^{\sigma - 1} du \leq \frac{k_1(\sigma)}{\sigma - \sigma_1},$$

and then by (5.3), it follows that

$$\frac{k_1(\sigma)}{\sigma - \sigma_1} \geq I_1 \geq M_1 J_1 = M_1 n. \tag{5.4}$$

By (5.4), setting $n \to \infty$, in view of $k_1(\sigma) < \infty$, $\sigma - \sigma_1 > 0$ and $M_1 > 0$, we find that

$$\infty > \frac{k_1(\sigma)}{\sigma - \sigma_1} \geq \infty,$$

which is a contradiction.

If $\sigma_1 > \sigma$, then for

$$n \geq \frac{1}{|q|(\sigma_1 - \sigma)} \quad (n \in \mathbf{N}),$$

we set the following two functions:

$$\tilde{f}_n(x) = \begin{cases} 0, 0 < x < 1 \\ x^{\sigma - \frac{1}{pn} - 1}, x \geq 1 \end{cases},$$

$$\tilde{g}_n(y) = \begin{cases} y^{\sigma_1 + \frac{1}{qn} - 1}, 0 < y \leq 1 \\ 0, y > 1 \end{cases},$$

and get that

$$\tilde{J}_1 = \left[\int_0^\infty x^{p(1-\sigma)-1} \tilde{f}_n^p(x) dx \right]^{\frac{1}{p}} \left[\int_0^\infty y^{q(1-\sigma_1)-1} \tilde{g}_n^q(y) dy \right]^{\frac{1}{q}}$$

$$= \left(\int_1^\infty x^{-\frac{1}{n}-1} dx \right)^{\frac{1}{p}} \left(\int_0^1 y^{\frac{1}{n}-1} dy \right)^{\frac{1}{q}} = n.$$

For fixed $x > 0$, setting $u = xy$, in view of $\sigma_1 + \frac{1}{qn} \geq \sigma$, we obtain

$$\tilde{I}_1 = \int_0^\infty \tilde{f}_n(x) \left(\int_0^{\frac{1}{x}} h(xy) \tilde{g}_n(y) dy \right) dx$$

$$= \int_1^\infty \left(\int_0^{\frac{1}{x}} h(xy) y^{\sigma_1 + \frac{1}{qn} - 1} dy \right) x^{\sigma - \frac{1}{pn} - 1} dx$$

$$= \int_1^\infty x^{(\sigma - \sigma_1) - \frac{1}{n} - 1} dx \int_0^1 h(u) u^{\sigma_1 + \frac{1}{qn} - 1} du$$

$$\leq \frac{1}{\sigma_1 - \sigma + \frac{1}{n}} \int_0^1 h(u) u^{\sigma - 1} du \leq \frac{k_1(\sigma)}{\sigma_1 - \sigma},$$

and then by Fubini's theorem (cf. [1]) and (5.3), we derive that

$$\frac{k_1(\sigma)}{\sigma_1 - \sigma} \geq \tilde{I}_1 = \int_0^\infty \tilde{g}_n(y) \left(\int_0^{\frac{1}{y}} h(xy) \tilde{f}_n(x) dx \right) dy$$

$$\geq M_1 \tilde{J}_1 = M_1 n. \tag{5.5}$$

By (5.5), setting $n \to \infty$, we see that

$$\infty > \frac{k_1(\sigma)}{\sigma_1 - \sigma} \geq \infty,$$

which is a contradiction.

Hence, we conclude that $\sigma_1 = \sigma$.

For $\sigma_1 = \sigma$, we deduce $I_1 \geq M_1 J_1$ and then it follows that

$$k_1(\sigma) = \int_0^1 h(u)u^{\sigma-1}du \geq \int_0^1 h(u)u^{\sigma+\frac{1}{pn}-1}du \geq M_1(>0). \qquad (5.6)$$

This completes the proof of the lemma.   □

Lemma 5.2 *If $k_2(\sigma) < \infty$, and if there exists a constant $M_2 > 0$, such that for any nonnegative measurable functions $f(x)$ and $g(y)$ in $(0, \infty)$, the following inequality*

$$\int_0^\infty g(y) \left(\int_{\frac{1}{y}}^\infty h(xy)f(x)dx \right) dy$$

$$\geq M_2 \left[\int_0^\infty x^{p(1-\sigma)-1} f^p(x)dx \right]^{\frac{1}{p}} \left[\int_0^\infty y^{q(1-\sigma_1)-1} g^q(y)dy \right]^{\frac{1}{q}} \qquad (5.7)$$

holds true, then we have

$$\sigma_1 = \sigma \quad and \quad k_2(\sigma) \geq M_2 \quad (> 0).$$

Proof If $\sigma_1 > \sigma$, then for $n \in \mathbf{N}$, we consider two functions $\widetilde{f}_n(x)$ and $\widetilde{g}_n(y)$ as in Lemma 5.1 and get that

$$\widetilde{J}_1 = \left[\int_0^\infty x^{p(1-\sigma)-1} \widetilde{f}_n^p(x)dx \right]^{\frac{1}{p}} \left[\int_0^\infty y^{q(1-\sigma_1)-1} \widetilde{g}_n^q(y)dy \right]^{\frac{1}{q}} = n.$$

For fixed $y > 0$, setting $u = xy$, we obtain

$$\widetilde{I}_2 = \int_0^\infty \widetilde{g}_n(y) \left(\int_{\frac{1}{y}}^\infty h(xy)\widetilde{f}_n(x)dx \right) dy$$

$$= \int_0^1 \left(\int_{\frac{1}{y}}^\infty h(xy)x^{\sigma-\frac{1}{pn}-1}dx \right) y^{\sigma_1+\frac{1}{qn}-1}dy$$

$$= \int_0^1 y^{(\sigma_1-\sigma)+\frac{1}{n}-1}dy \int_1^\infty h(u)u^{\sigma-\frac{1}{pn}-1}du \leq \frac{k_2(\sigma)}{\sigma_1 - \sigma},$$

and then by (5.7), it follows that

$$\frac{k_2(\sigma)}{\sigma_1 - \sigma} \geq \widetilde{I}_2 \geq M_2\widetilde{J}_1 = M_2 n. \qquad (5.8)$$

By (5.8), setting $n \to \infty$, we derive that

$$\infty > \frac{k_2(\sigma)}{\sigma_1 - \sigma} \geq \infty,$$

which is a contradiction.

If $\sigma_1 < \sigma$, then for

$$n \geq \frac{1}{|q|(\sigma - \sigma_1)} \quad (n \in \mathbf{N}),$$

we set two functions $f_n(x)$ and $g_n(y)$ as in Lemma 5.1 and get

$$J_1 = \left[\int_0^\infty x^{p(1-\sigma)-1} f_n^p(x) dx \right]^{\frac{1}{p}} \left[\int_0^\infty y^{q(1-\sigma_1)-1} g_n^q(y) dy \right]^{\frac{1}{q}} = n.$$

For fixed $x > 0$, setting $u = xy$, we obtain

$$I_2 = \int_0^\infty f_n(x) \left(\int_{\frac{1}{x}}^\infty h(xy) g_n(y) dy \right) dx$$

$$= \int_0^1 \left(\int_{\frac{1}{x}}^\infty h(xy) y^{\sigma_1 - \frac{1}{qn} - 1} dy \right) x^{\sigma + \frac{1}{pn} - 1} dx$$

$$= \int_0^1 x^{(\sigma - \sigma_1) + \frac{1}{n} - 1} dx \int_1^\infty h(u) u^{\sigma_1 - \frac{1}{qn} - 1} du \leq \frac{k_2(\sigma)}{\sigma - \sigma_1},$$

and then by Fubini's theorem (cf. [1]) and (5.7), it follows that

$$\frac{k_2(\sigma)}{\sigma - \sigma_1} \geq I_2 = \int_0^\infty g_n(y) \left(\int_{\frac{1}{y}}^\infty h(xy) f_n(x) dx \right) dy$$

$$\geq M_2 J_1 = M_2 n. \tag{5.9}$$

By (5.9), setting $n \to \infty$, it follows that

$$\infty > \frac{k_2(\sigma)}{\sigma - \sigma_1} \geq \infty,$$

which is a contradiction.

Hence, we conclude that $\sigma_1 = \sigma$.

For $\sigma_1 = \sigma$, we deduce that

$$\tilde{I}_2 \geq M_2 \tilde{J}_2$$

and thus it follows that

$$k_2(\sigma) = \int_1^\infty h(u) u^{\sigma - 1} du \geq \int_1^\infty h(u) u^{\sigma - \frac{1}{pn} - 1} \geq M_2 \ (> 0). \tag{5.10}$$

This completes the proof of the lemma. $\qquad\square$

5.2   Reverse Hardy-Type Integral Inequalities of the First Kind

Theorem 5.3 *Let M_1 be a positive constant. If $k_1(\sigma) < \infty$, then the following statements (i), (ii) and (iii) are equivalent:*

(i) For any $f(x) \geq 0$, satisfying

$$0 < \int_0^\infty x^{p(1-\sigma)-1} f^p(x) dx < \infty,$$

we have the following reverse Hardy-type integral inequality of the first kind with the nonhomogeneous kernel:

$$J := \left[\int_0^\infty y^{p\sigma_1 - 1} \left(\int_0^{\frac{1}{y}} h(xy) f(x) dx \right)^p dy \right]^{\frac{1}{p}}$$

$$> M_1 \left[\int_0^\infty x^{p(1-\sigma)-1} f^p(x) dx \right]^{\frac{1}{p}}. \tag{5.11}$$

(ii) For any $g(y) \geq 0$, satisfying

$$0 < \int_0^\infty y^{q(1-\sigma_1)-1} g^q(y) dy < \infty,$$

we have the following reverse Hardy-type integral inequality:

$$L_1 := \left[\int_0^\infty x^{q\sigma - 1} \left(\int_0^{\frac{1}{y}} h(xy) g(y) dy \right)^q dx \right]^{\frac{1}{q}}$$

$$> M_1 \left[\int_0^\infty y^{q(1-\sigma_1)-1} g^q(y) dy \right]^{\frac{1}{q}}. \tag{5.12}$$

(iii) For any $f(x) \geq 0$, satisfying

$$0 < \int_0^\infty x^{p(1-\sigma)-1} f^p(x) dx < \infty,$$

and $g(y) \geq 0$, satisfying

$$0 < \int_0^\infty y^{q(1-\sigma_1)-1} g^q(y) dy < \infty,$$

we have the following inequality:

$$I := \int_0^\infty g(y) \left(\int_0^{\frac{1}{y}} h(xy) f(x) dx \right) dy$$

$$> M_1 \left[\int_0^\infty x^{p(1-\sigma)-1} f^p(x) dx \right]^{\frac{1}{p}} \left[\int_0^\infty y^{q(1-\sigma_1)-1} g^q(y) dy \right]^{\frac{1}{q}}. \quad (5.13)$$

(iv) $\sigma_1 = \sigma$, and $k_1(\sigma) \geq M_1 (> 0)$.

If statement (iv) holds true, then the constant $M_1 = k_1(\sigma) (\in \mathbf{R}_+)$ in (5.11), (5.12) and (5.13) is the best possible.

Proof $(i) \Rightarrow (iii)$. By the reverse Hölder inequality (cf. [2]), we have

$$I = \int_0^\infty \left(y^{\sigma_1 - \frac{1}{p}} \int_0^{\frac{1}{y}} h(xy) f(x) dx \right) \left(y^{\frac{1}{p} - \sigma_1} g(y) \right) dy$$

$$\geq J \left[\int_0^\infty y^{q(1-\sigma_1)-1} g^q(y) dy \right]^{\frac{1}{q}}. \quad (5.14)$$

Then by (5.11), we have (5.12).

$(iii) \Rightarrow (iv)$. By Lemma 5.1, we have $\sigma_1 = \sigma$, and $k_1(\sigma) \geq M_1 (> 0)$.

$(iv) \Rightarrow (i)$. For fixed $y > 0$, setting $u = xy$, we obtain the following weight function:

$$\omega_1(\sigma, y) := y^\sigma \int_0^{\frac{1}{y}} h(xy) x^{\sigma-1} dx$$

$$= \int_0^1 h(u) u^{\sigma-1} du = k_1(\sigma) \ (y > 0).$$

By the reverse Hölder inequality with weight, for $y \in (0, \infty)$, we have

$$\left(\int_0^{\frac{1}{y}} h(xy) f(x) dx \right)^p$$

$$= \left\{ \int_0^{\frac{1}{y}} h(xy) \left[\frac{y^{(\sigma-1)/p}}{x^{(\sigma-1)/q}} f(x) \right] \left[\frac{x^{(\sigma-1)/q}}{y^{(\sigma-1)/p}} \right] dx \right\}^p$$

$$\geq \int_0^{\frac{1}{y}} h(xy) \frac{y^{\sigma-1}}{x^{(\sigma-1)p/q}} f^p(x) dx \left[\int_0^{\frac{1}{y}} h(xy) \frac{x^{\sigma-1}}{y^{(\sigma-1)q/p}} dx \right]^{p-1}$$

$$= \left[\omega_1(\sigma, y) y^{q(1-\sigma)-1} \right]^{p-1} \int_0^{\frac{1}{y}} h(xy) \frac{y^{\sigma-1}}{x^{(\sigma-1)p/q}} f^p(x) dx$$

$$= (k_1(\sigma))^{p-1} y^{-p\sigma+1} \int_0^{\frac{1}{y}} h(xy) \frac{y^{\sigma-1}}{x^{(\sigma-1)p/q}} f^p(x) dx. \quad (5.15)$$

If (5.15) assumes the form of equality for some $y \in (0, \infty)$, then (cf. [2]) there exist constants A and B, such that they are not all zero, and

$$A \frac{y^{\sigma-1}}{x^{(\sigma-1)p/q}} f^p(x) = B \frac{x^{\sigma-1}}{y^{(\sigma-1)q/p}} \quad a.e. \text{ in } \mathbf{R}_+.$$

Let us suppose that $A \neq 0$ (otherwise $B = A = 0$). It follows that

$$x^{p(1-\sigma)-1} f^p(x) = y^{q(1-\sigma)} \frac{B}{Ax} \quad a.e. \text{ in } \mathbf{R}_+,$$

which contradicts the fact that

$$0 < \int_0^\infty x^{p(1-\sigma)-1} f^p(x) dx < \infty.$$

Hence, (5.15) assumes the form of strict inequality.

For $\sigma_1 = \sigma$, by Fubini's theorem (cf. [1]) and the above result, we have

$$J > (k_1(\sigma))^{\frac{1}{q}} \left\{ \int_0^\infty \left[\int_0^{\frac{1}{y}} h(xy) \frac{y^{\sigma-1}}{x^{(\sigma-1)p/q}} f^p(x) dx \right] dy \right\}^{\frac{1}{p}}$$

$$= (k_1(\sigma))^{\frac{1}{q}} \left\{ \int_0^\infty \left[\int_0^{\frac{1}{x}} h(xy) \frac{y^{\sigma-1}}{x^{(\sigma-1)(p-1)}} dy \right] f^p(x) dx \right\}^{\frac{1}{p}}$$

$$= (k_1(\sigma))^{\frac{1}{q}} \left[\int_0^\infty \omega_1(\sigma, x) x^{p(1-\sigma)-1} f^p(x) dx \right]^{\frac{1}{p}}$$

$$= k_1(\sigma) \left[\int_0^\infty x^{p(1-\sigma)-1} f^p(x) dx \right]^{\frac{1}{p}}.$$

Since $k_1(\sigma) \geq M_1 \, (> 0)$, (5.11) follows.

Therefore, the statements (i), (iii) and (iv) are equivalent.

$(ii) \Leftrightarrow (iii)$. By the reverse Hölder inequality, we have

$$I = \int_0^\infty \left(x^{\frac{1}{q}-\sigma} f(x) \right) \left(x^{\sigma-\frac{1}{q}} \int_0^{\frac{1}{y}} h(xy) g(y) dy \right) dx$$

$$\geq \left[\int_0^\infty x^{p(1-\sigma)-1} f^p(x) dx \right]^{\frac{1}{p}} L_1. \tag{5.16}$$

Then by (5.12), we have (5.13).

On the other hand, suppose that (5.13) is valid. We set

$$f(x) := x^{q\sigma-1}\left(\int_0^{\frac{1}{y}} h(xy)g(y)dy\right)^{q-1}, \quad x > 0.$$

If $L_1 = \infty$, then (5.12) is trivially valid; if $L_1 = 0$, then it is impossible. Suppose that $0 < L_1 < \infty$. By (5.13), we have

$$\infty > \int_0^\infty x^{p(1-\sigma)-1} f^p(x)dx = L_1^q = I$$

$$> M_1\left[\int_0^\infty x^{p(1-\sigma)-1} f^p(x)dx\right]^{\frac{1}{p}}\left[\int_0^\infty y^{q(1-\sigma_1)-1} g^q(y)dy\right]^{\frac{1}{q}} > 0,$$

$$L_1 = \left[\int_0^\infty x^{p(1-\sigma)-1} f^p(x)dx\right]^{\frac{1}{q}} > M_1\left[\int_0^\infty y^{q(1-\sigma_1)-1} g^q(y)dy\right]^{\frac{1}{q}},$$

namely, (5.12) follows.

Hence, statements (i), (ii), (iii) and (iv) are equivalent.

If the statement (iv) holds true, and if there exists a constant $M_1 \geq k_1(\sigma)$ such that (5.13) is satisfied, then we obtain that $k_1(\sigma) \geq M_1$. Hence, the constant factor $M_1 = k_1(\sigma)$ in (5.13) is the best possible.

The constant factor $M_1 = k_1(\sigma)$ ($\in \mathbf{R}_+$) in (5.11) (resp. (5.12)) is still the best possible. Otherwise, by (5.14) (for $\sigma_1 = \sigma$) (resp. (5.16)), we can conclude that the constant factor $M_1 = k_1(\sigma)$ in (5.13) is not the best possible.

This completes the proof of the theorem. $\qquad\square$

In particular, for $\sigma = \sigma_1 = \frac{1}{p}$ in Theorem 5.3, we derive the following corollary:

Corollary 5.4 *Let M_1 be a positive constant. If $k_1(\frac{1}{p}) < \infty$, then the following statements (i), (ii), (iii) and (iv) are equivalent:*

(i) *For any $f(x) \geq 0$, satisfying*

$$0 < \int_0^\infty x^{p-2} f^p(x)dx < \infty,$$

we have the following inequality:

$$\left[\int_0^\infty \left(\int_0^{\frac{1}{y}} h(xy)f(x)dx\right)^p dy\right]^{\frac{1}{p}} > M_1\left(\int_0^\infty x^{p-2} f^p(x)dx\right)^{\frac{1}{p}}. \tag{5.17}$$

(ii) For any $g(y) \geq 0$, satisfying

$$0 < \int_0^\infty g^q(y)dy < \infty,$$

we have the following inequality:

$$\left[\int_0^\infty x^{q-2}\left(\int_0^{\frac{1}{x}} h(xy)g(y)dy\right)^q dx\right]^{\frac{1}{q}} > M_1 \left(\int_0^\infty g^q(y)dy\right)^{\frac{1}{q}}. \tag{5.18}$$

(iii) For any $f(x) \geq 0$, satisfying

$$0 < \int_0^\infty x^{p-2}f^p(x)dx < \infty,$$

and $g(y) \geq 0$, satisfying

$$0 < \int_0^\infty g^q(y)dy < \infty,$$

we have the following inequality:

$$\int_0^\infty g(y)\left(\int_0^{\frac{1}{y}} h(xy)f(x)dx\right)dy$$

$$> M_1 \left(\int_0^\infty x^{p-2}f^p(x)dx\right)^{\frac{1}{p}}\left(\int_0^\infty g^q(y)dy\right)^{\frac{1}{q}}. \tag{5.19}$$

(iv) $k_1(\frac{1}{p}) \geq M_1 \ (> 0)$.

If statement (iv) holds true, then the constant $M_1 = k_1(\frac{1}{p}) \ (\in \mathbf{R}_+)$ in (5.17), (5.18) and (5.19) is the best possible.

Setting

$$y = \frac{1}{Y}, \quad G(Y) = g\left(\frac{1}{Y}\right)\frac{1}{Y^2}$$

in Theorem 5.3, and then replacing Y by y, we deduce the following corollary:

Corollary 5.5 *Let M_1 be a positive constant. If $k_1(\sigma) < \infty$, then the following statements (i), (ii), (iii) and (iv) are equivalent:*

(i) For any $f(x) \geq 0$, satisfying

$$0 < \int_0^\infty x^{p(1-\sigma)-1}f^p(x)dx < \infty,$$

we have the following inequality:

$$\left[\int_0^\infty y^{-p\sigma_1-1} \left(\int_0^y h\left(\frac{x}{y}\right) f(x)dx \right)^p dy \right]^{\frac{1}{p}}$$
$$> M_1 \left[\int_0^\infty x^{p(1-\sigma)-1} f^p(x)dx \right]^{\frac{1}{p}}. \tag{5.20}$$

(ii) For any $G(y) \geq 0$, satisfying

$$0 < \int_0^\infty y^{q(1+\sigma_1)-1} G^q(y)dy < \infty,$$

we have the following inequality:

$$\left[\int_0^\infty x^{q\sigma-1} \left(\int_0^x h\left(\frac{x}{y}\right) G(y)dy \right)^q dx \right]^{\frac{1}{q}}$$
$$> M_1 \left[\int_0^\infty y^{q(1+\sigma_1)-1} G^q(y)dy \right]^{\frac{1}{q}}. \tag{5.21}$$

(iii) For any $f(x) \geq 0$, satisfying

$$0 < \int_0^\infty x^{p(1-\sigma)-1} f^p(x)dx < \infty,$$

and $G(y) \geq 0$, satisfying

$$0 < \int_0^\infty y^{q(1+\sigma_1)-1} G^q(y)dy < \infty,$$

we have the following inequality:

$$\int_0^\infty G(y) \left(\int_0^y h\left(\frac{x}{y}\right) f(x)dx \right) dy$$
$$> M_1 \left[\int_0^\infty x^{p(1-\sigma)-1} f^p(x)dx \right]^{\frac{1}{p}} \left[\int_0^\infty y^{q(1+\sigma_1)-1} G^q(y)dy \right]^{\frac{1}{q}}. \tag{5.22}$$

(iv) $\sigma_1 = \sigma$, and $k_1(\sigma) \geq M_1 (> 0)$.

If statement (iv) holds true, then the constant $M_1 = k_1(\sigma) (\in \mathbf{R}_+)$ in (5.20), (5.21) and (5.22) is the best possible.

Note. $h(\frac{x}{y})$ is a homogeneous function of degree 0, namely,

$$h\left(\frac{x}{y}\right) = k_0(x, y).$$

Setting
$$h(u) = k_\lambda(u, 1),$$

where $k_\lambda(x, y)$ is a homogeneous function of degree $-\lambda$ ($\in \mathbf{R}$), for $g(y) = y^\lambda G(y)$ and $\mu_1 = \lambda - \sigma_1$ in Corollary 5.5, we obtain the following:

Corollary 5.6 *Let M_1 be a positive constant. If*

$$k_\lambda^{(1)}(\sigma) = \int_0^1 k_\lambda(u, 1)u^{\sigma-1}du < \infty,$$

then the following statements (i), (ii), (iii) and (iv) are equivalent:

(i) For any $f(x) \geq 0$, satisfying

$$0 < \int_0^\infty x^{p(1-\sigma)-1} f^p(x)dx < \infty,$$

we have the following reverse Hardy-type inequality of the first kind with a homogeneous kernel:

$$\left[\int_0^\infty y^{p\mu_1-1} \left(\int_0^y k_\lambda(x, y)f(x)dx\right)^p dy\right]^{\frac{1}{p}}$$
$$> M_1 \left[\int_0^\infty x^{p(1-\sigma)-1} f^p(x)dx\right]^{\frac{1}{p}}. \tag{5.23}$$

(ii) For any $g(y) \geq 0$, satisfying

$$0 < \int_0^\infty y^{q(1-\mu_1)-1} g^q(y)dy < \infty,$$

we have the following inequality:

$$\left[\int_0^\infty x^{q\sigma-1} \left(\int_0^x k_\lambda(x, y)g(y)dy\right)^q dx\right]^{\frac{1}{q}}$$
$$> M_1 \left[\int_0^\infty y^{q(1-\mu_1)-1} g^q(y)dy\right]^{\frac{1}{p}}. \tag{5.24}$$

(iii) For any $f(x) \geq 0$, satisfying

$$0 < \int_0^\infty x^{p(1-\sigma)-1} f^p(x)dx < \infty,$$

and $g(y) \geq 0$, satisfying

$$0 < \int_0^\infty y^{q(1-\mu_1)-1} g^q(y)dy < \infty,$$

we have the following inequality:

$$\int_0^\infty g(y) \left(\int_0^y k_\lambda(x, y)f(x)dx \right) dy$$
$$> M_1 \left[\int_0^\infty x^{p(1-\sigma)-1} f^p(x)dx \right]^{\frac{1}{p}} \left[\int_0^\infty y^{q(1-\mu_1)-1} g^q(y)dy \right]^{\frac{1}{q}}. \quad (5.25)$$

(iv) $\mu_1 = \mu$, and $k_\lambda^{(1)}(\sigma) \geq M_1 \, (> 0)$.

If statement (iv) holds true, then the constant $M_1 = k_\lambda^{(1)}(\sigma) (\in \mathbf{R}_+)$ in (5.23), (5.24) and (5.25) is the best possible.

In particular, for $\lambda = 1, \sigma = \frac{1}{q}, \mu_1 = \mu = \frac{1}{p}$ in Corollary 5.6, we obtain the following:

Corollary 5.7 *Let M_1 be a positive constant. If $k_1^{(1)}(\frac{1}{q}) < \infty$, then the following statements (i), (ii), (iii) and (iv) are equivalent:*

(i) For any $f(x) \geq 0$, satisfying

$$0 < \int_0^\infty f^p(x)dx < \infty,$$

we have the following inequality:

$$\left[\int_0^\infty \left(\int_0^y k_1(x, y)f(x)dx \right)^p dy \right]^{\frac{1}{p}} > M_1 \left(\int_0^\infty f^p(x)dx \right)^{\frac{1}{p}}. \quad (5.26)$$

(ii) For any $g(y) \geq 0$, satisfying

$$0 < \int_0^\infty g^q(y)dy < \infty,$$

we have the following inequality:

$$\left[\int_0^\infty \left(\int_0^x k_1(x, y)g(y)dy\right)^q dx\right]^{\frac{1}{q}} > M_1 \left(\int_0^\infty g^q(y)dy\right)^{\frac{1}{p}}. \qquad (5.27)$$

(iii) For any $f(x) \geq 0$, satisfying

$$0 < \int_0^\infty f^p(x)dx < \infty,$$

and $g(y) \geq 0$, satisfying

$$0 < \int_0^\infty g^q(y)dy < \infty,$$

we have the following inequality:

$$I = \int_0^\infty g(y) \left(\int_0^y k_1(x, y)f(x)dx\right) dy$$

$$> M_1 \left(\int_0^\infty f^p(x)dx\right)^{\frac{1}{p}} \left(\int_0^\infty g^q(y)dy\right)^{\frac{1}{q}}. \qquad (5.28)$$

(iv) $k_1^{(1)}(\frac{1}{q}) \geq M_1 \ (> 0)$.

If statement (iv) holds true, then the constant $M_1 = k_1^{(1)}(\frac{1}{q}) \ (\in \mathbf{R}_+)$ in (5.26), (5.27) and (5.28) is the best possible.

5.3 Reverse Hardy-Type Inequalities of the Second Kind

Similarly, we obtain the following weight function:

$$\omega_2(\sigma, y) := y^\sigma \int_{\frac{1}{y}}^\infty h(xy)x^{\sigma-1}dx$$

$$= \int_1^\infty h(u)u^{\sigma-1}du = k_2(\sigma)(y > 0),$$

in view of Lemma 5.2, we similarly obtain the following theorem:

Theorem 5.8 *Let M_2 be a positive constant. If $k_2(\sigma) < \infty$, then the following statements (i), (ii), (iii) and (iv) are equivalent:*

(i) For any $f(x) \geq 0$, satisfying

$$0 < \int_0^\infty x^{p(1-\sigma)-1}f^p(x)dx < \infty,$$

we have the following reverse Hardy-type inequality of the second kind with a non-homogeneous kernel:

$$\left[\int_0^\infty y^{p\sigma_1-1}\left(\int_{\frac{1}{y}}^\infty h(xy)f(x)dx\right)^p dy\right]^{\frac{1}{p}}$$

$$> M_2\left[\int_0^\infty x^{p(1-\sigma)-1}f^p(x)dx\right]^{\frac{1}{p}}. \tag{5.29}$$

(ii) For any $g(y) \geq 0$, satisfying

$$0 < \int_0^\infty y^{q(1-\sigma_1)-1}g^q(y)dy < \infty,$$

we have the following inequality:

$$\left[\int_0^\infty x^{q\sigma-1}\left(\int_{\frac{1}{y}}^\infty h(xy)g(y)dy\right)^q dx\right]^{\frac{1}{q}}$$

$$> M_2\left[\int_0^\infty y^{q(1-\sigma_1)-1}g^q(y)dy\right]^{\frac{1}{q}}. \tag{5.30}$$

(iii) For any $f(x) \geq 0$, satisfying

$$0 < \int_0^\infty x^{p(1-\sigma)-1}f^p(x)dx < \infty,$$

and $g(y) \geq 0$, satisfying

$$0 < \int_0^\infty y^{q(1-\sigma_1)-1}g^q(y)dy < \infty,$$

we have the following inequality:

$$\int_0^\infty g(y)\left(\int_{\frac{1}{y}}^\infty h(xy)f(x)dx\right)dy$$

$$> M_2\left[\int_0^\infty x^{p(1-\sigma)-1}f^p(x)dx\right]^{\frac{1}{p}}\left[\int_0^\infty y^{q(1-\sigma_1)-1}g^q(y)dy\right]^{\frac{1}{q}}. \tag{5.31}$$

(iv) $\sigma_1 = \sigma$, and $k_2(\sigma) \geq M_2 (> 0)$.

If statement (iv) holds true, then the constant $M_2 = k_2(\sigma) (\in \mathbf{R}_+)$ in (5.29), (5.30) and (5.31) is the best possible.

In particular, for $\sigma = \sigma_1 = \frac{1}{p}$ in Theorem 5.8, we deduce the corollary below:

Corollary 5.9 *Let M_2 be a positive constant. If $k_2(\frac{1}{p}) < \infty$, then the following statements (i), (ii), (iii) and (iv) are equivalent:*

(i) For any $f(x) \geq 0$, satisfying

$$0 < \int_0^\infty x^{p-2} f^p(x) dx < \infty,$$

we have the following inequality:

$$\left[\int_0^\infty \left(\int_{\frac{1}{y}}^\infty h(xy) f(x) dx \right)^p dy \right]^{\frac{1}{p}} > M_2 \left(\int_0^\infty x^{p-2} f^p(x) dx \right)^{\frac{1}{p}}. \qquad (5.32)$$

(ii) For any $g(y) \geq 0$, satisfying

$$0 < \int_0^\infty g^q(y) dy < \infty,$$

we have the following inequality:

$$\left[\int_0^\infty x^{q-2} \left(\int_{\frac{1}{x}}^\infty h(xy) g(y) dy \right)^q dx \right]^{\frac{1}{q}} > M_2 \left[\int_0^\infty g^q(y) dy \right]^{\frac{1}{q}}. \qquad (5.33)$$

(iii) For any $f(x) \geq 0$, satisfying

$$0 < \int_0^\infty x^{p-2} f^p(x) dx < \infty,$$

and $g(y) \geq 0$, satisfying

$$0 < \int_0^\infty g^q(y) dy < \infty,$$

we have the following inequality:

$$\int_0^\infty g(y) \left(\int_{\frac{1}{y}}^\infty h(xy) f(x) dx \right) dy$$

$$> M_2 \left(\int_0^\infty x^{p-2} f^p(x) dx \right)^{\frac{1}{p}} \left(\int_0^\infty g^q(y) dy \right)^{\frac{1}{q}}. \qquad (5.34)$$

(iv) $k_2(\frac{1}{p}) \geq M_2 \ (> 0)$.

If statement (iv) holds true, then the constant $M_2 = k_2(\frac{1}{p}) \ (\in \mathbf{R}_+)$ in (5.32), (5.33) and (5.34) is the best possible.

Setting

$$y = \frac{1}{Y}, \quad G(Y) = g\left(\frac{1}{Y}\right) \frac{1}{Y^2}$$

in Corollary 5.9, and then replacing Y by y, we get the following corollary:

Corollary 5.10 *Let M_2 be a positive constant. If $k_2(\sigma) < \infty$, then the following statements (i), (ii), (iii) and (iv) are equivalent:*

(i) For any $f(x) \geq 0$, satisfying

$$0 < \int_0^\infty x^{p(1-\sigma)-1} f^p(x)dx < \infty,$$

we have the following inequality:

$$\left[\int_0^\infty y^{-p\sigma_1-1} \left(\int_y^\infty h\left(\frac{x}{y}\right) f(x)dx \right)^p dy \right]^{\frac{1}{p}}$$

$$> M_2 \left[\int_0^\infty x^{p(1-\sigma)-1} f^p(x)dx \right]^{\frac{1}{p}}. \tag{5.35}$$

(ii) For any $G(y) \geq 0$, satisfying

$$0 < \int_0^\infty y^{q(1+\sigma_1)-1} G^q(y)dy < \infty,$$

we have the following inequality:

$$\left[\int_0^\infty x^{q\sigma-1} \left(\int_x^\infty h\left(\frac{x}{y}\right) G(y)dy \right)^q dx \right]^{\frac{1}{q}}$$

$$> M_2 \left[\int_0^\infty y^{q(1+\sigma_1)-1} G^q(y)dy \right]^{\frac{1}{q}}. \tag{5.36}$$

(iii) For any $f(x), G(y) \geq 0$, satisfying

$$0 < \int_0^\infty x^{p(1-\sigma)-1} f^p(x)dx < \infty,$$

and $G(y) \geq 0$, satisfying

$$0 < \int_0^\infty y^{q(1+\sigma_1)-1} G^q(y)dy < \infty,$$

we have the following inequality:

$$\int_0^\infty G(y) \left(\int_y^\infty h\left(\frac{x}{y}\right) f(x)dx \right) dy$$

$$> M_2 \left[\int_0^\infty x^{p(1-\sigma)-1} f^p(x)dx \right]^{\frac{1}{p}} \left[\int_0^\infty y^{q(1+\sigma_1)-1} G^q(y)dy \right]^{\frac{1}{q}}. \tag{5.37}$$

(iv) $\sigma_1 = \sigma$, and $k_2(\sigma) \geq M_2 \ (> 0)$.

If statement (iv) holds true, then the constant $M_2 = k_2(\sigma) \ (\in \mathbf{R}_+)$ in (5.35), (5.36) and (5.37) is the best possible.

Setting

$$h(u) = k_\lambda(u, 1),$$

where $k_\lambda(x, y)$ is a homogeneous function of degree $-\lambda \ (\in \mathbf{R})$, for $g(y) = y^\lambda G(y)$ and $\mu_1 = \lambda - \sigma_1$ in Corollary 5.10, we get the following:

Corollary 5.11 *Let M_2 be a positive constant. If*

$$k_\lambda^{(2)}(\sigma) = \int_1^\infty k_\lambda(u, 1)u^{\sigma-1}du < \infty,$$

then the following statements (i), (ii), (iii) and (iv) are equivalent:

(i) For any $f(x) \geq 0$, satisfying

$$0 < \int_0^\infty x^{p(1-\sigma)-1} f^p(x)dx < \infty,$$

we have the following reverse Hardy-type integral inequality of the second kind with a homogeneous kernel:

$$\left[\int_0^\infty y^{p\mu_1-1} \left(\int_y^\infty k_\lambda(x, y)f(x)dx \right)^p dy \right]^{\frac{1}{p}}$$
$$> M_2 \left[\int_0^\infty x^{p(1-\sigma)-1} f^p(x)dx \right]^{\frac{1}{p}}. \tag{5.38}$$

(ii) For any $g(y) \geq 0$, satisfying

$$0 < \int_0^\infty y^{q(1-\mu_1)-1} g^q(y)dy < \infty,$$

we have the following inequality:

$$\left[\int_0^\infty x^{q\sigma-1} \left(\int_x^\infty k_\lambda(x, y)g(y)dy \right)^q dx \right]^{\frac{1}{q}}$$
$$> M_2 \left[\int_0^\infty y^{q(1-\mu_1)-1} g^q(y)dy \right]^{\frac{1}{p}}. \tag{5.39}$$

(iii) For any $f(x) \geq 0$, satisfying

$$0 < \int_0^\infty x^{p(1-\sigma)-1} f^p(x)dx < \infty,$$

and $g(y) \geq 0$, satisfying

$$0 < \int_0^\infty y^{q(1-\mu_1)-1} g^q(y)dy < \infty,$$

we have the following inequality:

$$\int_0^\infty g(y) \left(\int_y^\infty k_\lambda(x, y) f(x)dx \right) dy$$

$$> M_2 \left[\int_0^\infty x^{p(1-\sigma)-1} f^p(x)dx \right]^{\frac{1}{p}} \left[\int_0^\infty y^{q(1-\mu_1)-1} g^q(y)dy \right]^{\frac{1}{q}}. \quad (5.40)$$

(iv) $\mu_1 = \mu$, and $k_\lambda^{(2)}(\sigma) \geq M_2 \ (> 0)$.

If statement (iv) holds true, then the constant $M_2 = k_\lambda^{(2)}(\sigma) \ (\in \mathbf{R}_+)$ in (5.38), (5.39) and (5.40) is the best possible.

In particular, for $\lambda = 1, \sigma = \frac{1}{q}, \mu = \frac{1}{p}$ in Corollary 5.11, we deduce the corollary below:

Corollary 5.12 *Let M_2 be a positive constant. If $k_1^{(2)}(\frac{1}{q}) < \infty$, then the following statements (i), (ii), (iii) and (iv) are equivalent:*

(i) For any $f(x) \geq 0$, satisfying

$$0 < \int_0^\infty f^p(x)dx < \infty,$$

we have the following inequality:

$$\left[\int_0^\infty \left(\int_y^\infty k_1(x, y) f(x)dx \right)^p dy \right]^{\frac{1}{p}} > M_2 \left(\int_0^\infty f^p(x)dx \right)^{\frac{1}{p}}. \quad (5.41)$$

(ii) For any $g(y) \geq 0$, satisfying

$$0 < \int_0^\infty g^q(y)dy < \infty,$$

we have the following inequality:

$$\left[\int_0^\infty \left(\int_x^\infty k_1(x, y) g(y)dy \right)^q dx \right]^{\frac{1}{q}} > M_2 \left(\int_0^\infty g^q(y)dy \right)^{\frac{1}{p}}. \quad (5.42)$$

(iii) For any $f(x) \geq 0$, satisfying

$$0 < \int_0^\infty f^p(x)dx < \infty,$$

and $g(y) \geq 0$, satisfying

$$0 < \int_0^\infty g^q(y)dy < \infty,$$

we have the following inequality:

$$\int_0^\infty g(y)\left(\int_y^\infty k_1(x, y)f(x)dx\right)dy$$

$$> M_2 \left(\int_0^\infty f^p(x)dx\right)^{\frac{1}{p}}\left(\int_0^\infty g^q(y)dy\right)^{\frac{1}{q}}. \tag{5.43}$$

(iv) $k_1^{(2)}(\frac{1}{q}) \geq M_2$ (> 0).

If statement (iv) holds true, then the constant $M_2 = k_1^{(2)}(\frac{1}{q})$ $(\in \mathbf{R}_+)$ in (5.41), (5.42) and (5.43) is the best possible.

Example 5.13 Setting

$$h(u) = k_\lambda(u, 1) = \frac{|\ln u|^\beta (\max\{u, 1\})^\alpha}{|u^\lambda - 1|(\min\{u, 1\})^\alpha} \quad (u > 0),$$

then we obtain that

$$h(xy) = \frac{|\ln xy|^\beta (\max\{xy, 1\})^\alpha}{|(xy)^\lambda - 1|(\min\{xy, 1\})^\alpha},$$

$$k_\lambda(x, y) = \frac{|\ln x/y|^\beta (\max\{x, y\})^\alpha}{|x^\lambda - y^\lambda|(\min\{x, y\})^\alpha},$$

and for $\beta, \lambda > 0, \sigma > \alpha$,

$$k_1(\sigma) = k_\lambda^{(1)}(\sigma) = \int_0^1 \frac{(-\ln u)^\beta}{1 - u^\lambda} u^{\sigma-\alpha-1}du$$

$$= \int_0^1 (-\ln u)^\beta \sum_{k=0}^\infty u^{k\lambda+\sigma-\alpha-1}du$$

$$= \sum_{k=0}^\infty \int_0^1 (-\ln u)^\beta u^{k\lambda+\sigma-\alpha-1}du.$$

Setting $v = (k\lambda + \sigma - \alpha)(-\ln u)$ in the previous integral, we have

$$k_1(\sigma) = k_\lambda^{(1)}(\sigma) = \sum_{k=0}^{\infty} \frac{1}{(k\lambda + \sigma - \alpha)^{\beta+1}} \int_0^\infty v^\beta e^{-v} dv$$

$$= \frac{\Gamma(\beta+1)}{\lambda^{\beta+1}} \zeta\left(\beta + 1, \frac{\sigma - \alpha}{\lambda}\right) \in \mathbf{R}_+, \qquad (5.44)$$

where

$$\Gamma(\eta) := \int_0^\infty v^{\eta-1} e^{-v} dv \ (\eta > 0)$$

is the gamma function and

$$\zeta(s, a) := \sum_{k=0}^{\infty} \frac{1}{(k+a)^s} (Res > 1, a > 0)$$

is the extended Hurwitz zeta function (cf. [3]).

For $\beta, \lambda > 0, \mu = \lambda - \sigma > \alpha$, setting $v = \frac{1}{u}$, we get

$$k_2(\sigma) = k_\lambda^{(2)}(\sigma) = \int_1^\infty \frac{(\ln u)^\beta}{u^\lambda - 1} u^{\sigma+\alpha-1} du$$

$$= \int_0^1 \frac{(-\ln v)^\beta}{1 - v^\lambda} u^{\mu-\alpha-1} du$$

$$= \frac{\Gamma(\beta+1)}{\lambda^{\beta+1}} \zeta\left(\beta + 1, \frac{\mu - \alpha}{\lambda}\right) \in \mathbf{R}_+. \qquad (5.45)$$

Note. If $\beta, \lambda > 0, \sigma, \mu > -\alpha$, then $\lambda = \sigma + \mu > -2\alpha$, and $k_1(\sigma), k_2(\sigma) \in \mathbf{R}_+$.

Example 5.14 Setting

$$h(u) = k_\lambda(u, 1) = \frac{|\ln u|^\beta (\min\{u, 1\})^{\alpha+\gamma}}{|u^\alpha - 1|(\max\{u, 1\})^{\lambda+\gamma}} (u > 0),$$

then we get

$$h(xy) = \frac{|\ln xy|^\beta (\min\{xy, 1\})^{\alpha+\gamma}}{|(xy)^\alpha - 1|(\max\{xy, 1\})^{\lambda+\gamma}},$$

$$k_\lambda(x, y) = \frac{|\ln x/y|^\beta (\min\{x, y\})^{\alpha+\gamma}}{|x^\alpha - y^\alpha|(\max\{x, y\})^{\lambda+\gamma}},$$

and for $\beta, \alpha > 0, \sigma > -\alpha - \gamma$,

$$k_1(\sigma) = k_\lambda^{(1)}(\sigma) = \int_0^1 \frac{(-\ln u)^\beta}{1 - u^\alpha} u^{\sigma+\alpha+\gamma-1} du$$

$$= \int_0^1 (-\ln u)^\beta \sum_{k=0}^\infty u^{k\alpha+\sigma+\alpha+\gamma-1} du$$

$$= \sum_{k=0}^\infty \int_0^1 (-\ln u)^\beta u^{k\alpha+\sigma+\alpha+\gamma-1} du.$$

Setting $v = (k\alpha + \sigma + \alpha + \gamma)(-\ln u)$ in the previous integral, we have

$$k_1(\sigma) = k_\lambda^{(1)}(\sigma) = \sum_{k=0}^\infty \frac{1}{(k\alpha + \sigma + \alpha + \gamma)^{\beta+1}} \int_0^\infty v^\beta e^{-v} dv$$

$$= \frac{\Gamma(\beta+1)}{\alpha^{\beta+1}} \zeta\left(\beta+1, \frac{\sigma+\alpha+\gamma}{\alpha}\right) \in \mathbf{R}_+. \tag{5.46}$$

For $\beta, \alpha > 0, \mu > -\alpha - \gamma$, setting $v = \frac{1}{u}$, we find

$$k_2(\sigma) = k_\lambda^{(2)}(\sigma) = \int_1^\infty \frac{(\ln u)^\beta}{u^\alpha - 1} u^{\sigma-\lambda-\gamma-1} du$$

$$= \int_0^1 \frac{(-\ln v)^\beta}{1 - v^\alpha} v^{\mu+\alpha+\gamma-1} dv$$

$$= \frac{\Gamma(\beta+1)}{\alpha^{\beta+1}} \zeta\left(\beta+1, \frac{\mu+\alpha+\gamma}{\alpha}\right) \in \mathbf{R}_+. \tag{5.47}$$

Note. If $\beta > 0, \sigma, \mu > -\alpha - \gamma$, then $\lambda = \sigma + \mu > -2(\alpha + \gamma)$, and $k_1(\sigma) = k_\lambda^{(1)}(\sigma) \in \mathbf{R}_+$, $k_2(\sigma) = k_\lambda^{(2)}(\sigma) \in \mathbf{R}_+$.

Remark 5.15 We may use the above examples in order to construct two kinds of equivalent inequalities with particular kernels in Theorem 5.3, Corollary 5.6, Theorem 5.8 and Corollary 5.11.

5.4 Reverse Hardy-Type Integral Inequalities with the Interval Variable

For $a \in \mathbf{R} \backslash \{0\}, b > 0$, replacing x (resp. y) by e^{ax} (resp. e^{by}), then replacing $f(e^{ax})e^{ax}$ (resp. $g(e^{by})e^{by}$) by $f(x)$ (resp. $g(y)$), and $\frac{M}{|a|^{1/q}b^{1/p}}$ by M in Theorem 5.3, Corollary 5.6, Theorem 5.8 and Corollary 5.11, and carrying out the corresponding simplifications, we deduce the following theorem:

Theorem 5.16 *Let M_1 be a positive constant, and $a \neq 0, b > 0$. If $k_1(\sigma) < \infty$, then the following statements (i), (ii), (iii) and (iv) are equivalent:*

(i) For any $f(x) \geq 0$, satisfying

$$0 < \int_{-\infty}^{\infty} \left(\frac{f(x)}{e^{\sigma ax}} \right)^p dx < \infty,$$

we have the following inequality:

$$\left[\int_{-\infty}^{\infty} e^{p\sigma_1 by} \left(\int_{-\infty}^{e^{-by}} h(e^{ax+by}) f(x) dx \right)^p dy \right]^{\frac{1}{p}}$$

$$> M_1 \left[\int_{-\infty}^{\infty} \left(\frac{f(x)}{e^{\sigma ax}} \right)^p dx \right]^{\frac{1}{p}}. \tag{5.48}$$

(ii) For any $g(y) \geq 0$, satisfying

$$0 < \int_{-\infty}^{\infty} \left(\frac{g(y)}{e^{\sigma_1 by}} \right)^q dy < \infty,$$

we have the following reverse Hardy-type integral inequality:

$$\left[\int_{-\infty}^{\infty} e^{q\sigma ax} \left(\int_{-\infty}^{e^{-by}} h(e^{ax+by}) g(y) dy \right)^q dx \right]^{\frac{1}{q}}$$

$$> M_1 \left[\int_{-\infty}^{\infty} \left(\frac{g(y)}{e^{\sigma_1 by}} \right)^q dy dy \right]^{\frac{1}{q}}. \tag{5.49}$$

(iii) For any $f(x), g(y) \geq 0$, satisfying

$$0 < \int_{-\infty}^{\infty} \left(\frac{f(x)}{e^{\sigma ax}} \right)^p dx < \infty,$$

and $g(y) \geq 0$, satisfying

$$0 < \int_{-\infty}^{\infty} \left(\frac{g(y)}{e^{\sigma_1 by}} \right)^q dy < \infty,$$

we have the following inequality:

$$\int_{-\infty}^{\infty} g(y) \left(\int_{-\infty}^{e^{-by}} h(e^{ax+by}) f(x) dx \right) dy$$

$$> M_1 \left[\int_{-\infty}^{\infty} \left(\frac{f(x)}{e^{\sigma ax}} \right)^p dx \right]^{\frac{1}{p}} \left[\int_{-\infty}^{\infty} \left(\frac{g(y)}{e^{\sigma_1 by}} \right)^q dy \right]^{\frac{1}{q}}. \tag{5.50}$$

(iii)

$$\sigma_1 = \sigma \quad \text{and} \quad \frac{k_1(\sigma)}{|a|^{1/q}b^{1/p}} \geq M_1 \ (> 0).$$

If statement (iv) holds true, then the constant $M_1 = \frac{k_1(\sigma)}{|a|^{1/q}b^{1/p}}$ $(\in \mathbf{R}_+)$ in (5.48), (5.49) and (5.50) is the best possible.

In particular, for $\sigma_1 = \sigma$,

$$h(u) = \frac{|\ln u|^\beta (\max\{u, 1\})^\alpha}{|u^\lambda - 1|(\min\{u, 1\})^\alpha}$$

$(\beta, \lambda > 0, \sigma > \alpha)$, by Example 5.13, we have the following equivalent inequalities with

$$\frac{\Gamma(\beta + 1)}{a^{1/q}b^{1/p}\lambda^{\beta+1}} \zeta\left(\beta + 1, \frac{\sigma - \alpha}{\lambda}\right)$$

being the best possible constant factor:

$$\left\{ \int_{-\infty}^\infty e^{p\sigma by} \left[\int_{-\infty}^{e^{-by}} \frac{|\ln xy|^\beta (\max\{xy, 1\})^\alpha}{|(xy)^\lambda - 1|(\min\{xy, 1\})^\alpha} f(x)dx \right]^p dy \right\}^{\frac{1}{p}}$$
$$> \frac{\Gamma(\beta + 1)}{a^{1/q}b^{1/p}\lambda^{\beta+1}} \zeta\left(\beta + 1, \frac{\sigma - \alpha}{\lambda}\right) \left[\int_{-\infty}^\infty \left(\frac{f(x)}{e^{\sigma ax}}\right)^p dx \right]^{\frac{1}{p}}, \qquad (5.51)$$

$$\left[\int_{-\infty}^\infty e^{q\sigma ax} \left(\int_{-\infty}^{e^{-by}} \frac{|\ln xy|^\beta (\max\{xy, 1\})^\alpha}{|(xy)^\lambda - 1|(\min\{xy, 1\})^\alpha} g(y)dy \right)^q dx \right]^{\frac{1}{q}}$$
$$> \frac{\Gamma(\beta + 1)}{a^{1/q}b^{1/p}\lambda^{\beta+1}} \zeta\left(\beta + 1, \frac{\sigma - \alpha}{\lambda}\right) \left[\int_{-\infty}^\infty \left(\frac{g(y)}{e^{\sigma by}}\right)^q dy dy \right]^{\frac{1}{q}}, \qquad (5.52)$$

$$\int_{-\infty}^\infty g(y) \left[\int_{-\infty}^{e^{-by}} \frac{|\ln xy|^\beta (\max\{xy, 1\})^\alpha}{|(xy)^\lambda - 1|(\min\{xy, 1\})^\alpha} f(x)dx \right] dy$$
$$> \frac{\Gamma(\beta + 1)}{a^{1/q}b^{1/p}\lambda^{\beta+1}} \zeta\left(\beta + 1, \frac{\sigma - \alpha}{\lambda}\right)$$
$$\times \left[\int_{-\infty}^\infty \left(\frac{f(x)}{e^{\sigma ax}}\right)^p dx \right]^{\frac{1}{p}} \left[\int_{-\infty}^\infty \left(\frac{g(y)}{e^{\sigma by}}\right)^q dy \right]^{\frac{1}{q}}. \qquad (5.53)$$

Remark 5.17 In Theorem 5.16, if $\sigma_1 = \sigma$, $b < 0$, then replacing $-b$ by $b > 0$, we have the following equivalent inequalities with the best possible constant factor $\frac{k_1(\sigma)}{|a|^{1/q}b^{1/p}}$:

$$\left[\int_{-\infty}^{\infty} e^{-p\sigma by} \left(\int_{eby}^{\infty} h(e^{ax-by})f(x)dx\right)^p dy\right]^{\frac{1}{p}}$$

$$> \frac{k_1(\sigma)}{|a|^{1/q}b^{1/p}} \left[\int_{-\infty}^{\infty} \left(\frac{f(x)}{e^{\sigma ax}}\right)^p dx\right]^{\frac{1}{p}}. \tag{5.54}$$

$$\left[\int_{-\infty}^{\infty} e^{q\sigma ax} \left(\int_{eby}^{\infty} h(e^{ax-by})g(y)dy\right)^q dx\right]^{\frac{1}{q}}$$

$$> \frac{k_1(\sigma)}{|a|^{1/q}b^{1/p}} \left[\int_{-\infty}^{\infty} \left(\frac{g(y)}{e^{-\sigma by}}\right)^q dydy\right]^{\frac{1}{q}}. \tag{5.55}$$

$$\int_{-\infty}^{\infty} g(y) \left(\int_{eby}^{\infty} h(e^{ax-by})f(x)dx\right) dy$$

$$> \frac{k_1(\sigma)}{|a|^{1/q}b^{1/p}} \left[\int_{-\infty}^{\infty} \left(\frac{f(x)}{e^{\sigma ax}}\right)^p dx\right]^{\frac{1}{p}} \left[\int_{-\infty}^{\infty} \left(\frac{g(y)}{e^{-\sigma by}}\right)^q dy\right]^{\frac{1}{q}}. \tag{5.56}$$

In particular, for

$$h(u) = \frac{|\ln u|^\beta (\min\{u, 1\})^{\alpha+\gamma}}{|u^\alpha - 1|(\max\{u, 1\})^{\lambda+\gamma}}$$

$(\beta, \alpha > 0, \sigma > -\alpha - \gamma)$, by Example 5.14, we have the following equivalent inequalities with

$$\frac{\Gamma(\beta+1)}{|a|^{1/q}b^{1/p}\alpha^{\beta+1}} \zeta\left(\beta+1, \frac{\sigma+\alpha+\gamma}{\alpha}\right)$$

being the best possible constant factor:

$$\left\{\int_{-\infty}^{\infty} e^{-p\sigma by} \left[\int_{eby}^{\infty} \frac{|ax-by|^\beta (\min\{e^{ax-by}, 1\})^{\alpha+\gamma} f(x)}{|e^{ax-by} - 1|(\max\{e^{ax-by}, 1\})^\lambda} dx\right]^p dy\right\}^{\frac{1}{p}}$$

$$> \frac{\Gamma(\beta+1)}{|a|^{1/q}b^{1/p}\alpha^{\beta+1}} \zeta\left(\beta+1, \frac{\sigma+\alpha+\gamma}{\alpha}\right) \left[\int_{-\infty}^{\infty} \left(\frac{f(x)}{e^{\sigma ax}}\right)^p dx\right]^{\frac{1}{p}}, \tag{5.57}$$

$$\left[\int_{-\infty}^{\infty} e^{q\sigma ax} \left(\int_{eby}^{\infty} \frac{|ax-by|^\beta (\min\{e^{ax-by}, 1\})^{\alpha+\gamma} g(y)}{|e^{ax-by} - 1|(\max\{e^{ax-by}, 1\})^\lambda} dy\right)^q dx\right]^{\frac{1}{q}}$$

$$> \frac{\Gamma(\beta+1)}{|a|^{1/q}b^{1/p}\alpha^{\beta+1}} \zeta\left(\beta+1, \frac{\sigma+\alpha+\gamma}{\alpha}\right) \left[\int_{-\infty}^{\infty} \left(\frac{g(y)}{e^{-\sigma by}}\right)^q dydy\right]^{\frac{1}{q}}. \tag{5.58}$$

$$\int_{-\infty}^{\infty} g(y) \left[\int_{e^{by}}^{\infty} \frac{|ax - by|^{\beta} (\min\{e^{ax-by}, 1\})^{\alpha+\gamma}}{|e^{ax-by} - 1| (\max\{e^{ax-by}, 1\})^{\lambda}} f(x)dx \right] dy$$

$$> \frac{\Gamma(\beta+1)}{|a|^{1/q} b^{1/p} \alpha^{\beta+1}} \zeta\left(\beta+1, \frac{\sigma+\alpha+\gamma}{\alpha}\right)$$

$$\times \left[\int_{-\infty}^{\infty} \left(\frac{f(x)}{e^{\sigma ax}} \right)^{p} dx \right]^{\frac{1}{p}} \left[\int_{-\infty}^{\infty} \left(\frac{g(y)}{e^{-\sigma by}} \right)^{q} dy \right]^{\frac{1}{q}}. \tag{5.59}$$

Corollary 5.18 *Let M_1 be a positive constant, and a $\neq 0, b > 0$. If*

$$k_{\lambda}^{(1)}(\sigma) = \int_0^1 k_{\lambda}(u, 1) u^{\sigma-1} du < \infty,$$

then the following statements (i), (ii) (iii) and (iv) are equivalent:

(i) *For any $f(x) \geq 0$, satisfying*

$$0 < \int_{-\infty}^{\infty} \left(\frac{f(x)}{e^{\sigma ax}} \right)^{p} dx < \infty,$$

we have the following inequality:

$$\left[\int_{-\infty}^{\infty} e^{p\mu_1 by} \left(\int_{-\infty}^{e^{by}} k_{\lambda}(e^{ax}, e^{by}) f(x)dx \right)^{p} dy \right]^{\frac{1}{p}}$$

$$> M_1 \left[\int_{-\infty}^{\infty} \left(\frac{f(x)}{e^{\sigma ax}} \right)^{p} dx \right]^{\frac{1}{p}}. \tag{5.60}$$

(ii) *For any $g(y) \geq 0$, satisfying*

$$0 < \int_{-\infty}^{\infty} \left(\frac{g(y)}{e^{\mu_1 by}} \right)^{q} dy < \infty,$$

we have the following inequality:

$$\left[\int_{-\infty}^{\infty} e^{q\sigma ax} \left(\int_{-\infty}^{e^{by}} k_{\lambda}(e^{ax}, e^{by}) g(y)dy \right)^{q} dx \right]^{\frac{1}{q}}$$

$$> M_1 \left[\int_{-\infty}^{\infty} \left(\frac{g(y)}{e^{\mu_1 by}} \right)^{q} dy \right]^{\frac{1}{q}}. \tag{5.61}$$

(iii) For any $f(x) \geq 0$, satisfying

$$0 < \int_{-\infty}^{\infty} \left(\frac{f(x)}{e^{\sigma ax}}\right)^p dx < \infty,$$

and $g(y) \geq 0$, satisfying

$$0 < \int_{-\infty}^{\infty} \left(\frac{g(y)}{e^{\mu_1 by}}\right)^q dy < \infty,$$

we have the following inequality:

$$\int_{-\infty}^{\infty} g(y) \left(\int_{-\infty}^{e^{by}} k_\lambda(e^{ax}, e^{by}) f(x) dx\right) dy$$

$$> M_1 \left[\int_{-\infty}^{\infty} \left(\frac{f(x)}{e^{\sigma ax}}\right)^p dx\right]^{\frac{1}{p}} \left[\int_{-\infty}^{\infty} \left(\frac{g(y)}{e^{\mu_1 by}}\right)^q dy\right]^{\frac{1}{q}}. \tag{5.62}$$

(iii)

$$\mu_1 = \mu \quad \text{and} \quad \frac{k_\lambda^{(1)}(\sigma)}{|a|^{1/q}b^{1/p}} \geq M_1 \ (> 0).$$

If statement (iv) holds true, then the constant $M_1 = \frac{k_\lambda^{(1)}(\sigma)}{|a|^{1/q}b^{1/p}} \ (\in \mathbf{R}_+)$ in (5.60), (5.61) and (5.62) is the best possible.

In particular, for $\mu_1 = \mu$,

$$h(u) = \frac{|\ln u|^\beta (\max\{u, 1\})^\alpha}{|u^\lambda - 1|(\min\{u, 1\})^\alpha}$$

$(\beta, \lambda > 0, \sigma > \alpha)$, by Example 5.13, we have the following equivalent inequalities with

$$\frac{\Gamma(\beta + 1)}{a^{1/q}b^{1/p}\lambda^{\beta+1}} \zeta\left(\beta + 1, \frac{\sigma - \alpha}{\lambda}\right)$$

being the best possible constant factor:

$$\left\{\int_{-\infty}^{\infty} e^{p\mu by} \left[\int_{-\infty}^{e^{by}} \frac{|ax + by|^\beta (\min\{e^{ax}, e^{by}\})^\alpha f(x)}{|e^{\lambda ax} - e^{\lambda by}|(\max\{e^{ax}, e^{by}\})^\alpha} dx\right]^p dy\right\}^{\frac{1}{p}}$$

$$> \frac{\Gamma(\beta + 1)}{a^{1/q}b^{1/p}\lambda^{\beta+1}} \zeta\left(\beta + 1, \frac{\sigma - \alpha}{\lambda}\right) \left[\int_{-\infty}^{\infty} \left(\frac{f(x)}{e^{\sigma ax}}\right)^p dx\right]^{\frac{1}{p}}, \tag{5.63}$$

$$\left\{ \int_{-\infty}^{\infty} e^{q\sigma ax} \left[\int_{-\infty}^{e^{by}} \frac{|ax + by|^{\beta} (\min\{e^{ax}, e^{by}\})^{\alpha} g(y)}{|e^{\lambda ax} - e^{\lambda by}| (\max\{e^{ax}, e^{by}\})^{\alpha}} dy \right]^{q} dx \right\}^{\frac{1}{q}}$$

$$> \frac{\Gamma(\beta + 1)}{a|^{1/q} b^{1/p} \lambda^{\beta+1}} \zeta\left(\beta + 1, \frac{\sigma - \alpha}{\lambda}\right) \left[\int_{-\infty}^{\infty} \left(\frac{g(y)}{e^{\mu by}}\right)^{q} dy \right]^{\frac{1}{q}}, \qquad (5.64)$$

$$\int_{-\infty}^{\infty} g(y) \left[\int_{-\infty}^{e^{by}} \frac{|ax + by|^{\beta} (\min\{e^{ax}, e^{by}\})^{\alpha}}{|e^{\lambda ax} - e^{\lambda by}| (\max\{e^{ax}, e^{by}\})^{\alpha}} f(x) dx \right] dy$$

$$> \frac{\Gamma(\beta + 1)}{a|^{1/q} b^{1/p} \lambda^{\beta+1}} \zeta\left(\beta + 1, \frac{\sigma - \alpha}{\lambda}\right)$$

$$\times \left[\int_{-\infty}^{\infty} \left(\frac{f(x)}{e^{\sigma ax}}\right)^{p} dx \right]^{\frac{1}{p}} \left[\int_{-\infty}^{\infty} \left(\frac{g(y)}{e^{\mu by}}\right)^{q} dy \right]^{\frac{1}{q}}. \qquad (5.65)$$

Remark 5.19 In Corollary 5.18, if $\mu_1 = \mu, b < 0$, then replacing $-b$ by $b > 0$, we have the following equivalent inequalities with the best possible constant factor $\frac{k_{\lambda}^{(1)}(\sigma)}{|a|^{1/q} b^{1/p}}$:

$$\left[\int_{-\infty}^{\infty} e^{-p\mu by} \left(\int_{e^{-by}}^{\infty} k_{\lambda}(e^{ax}, e^{-by}) f(x) dx \right)^{p} dy \right]^{\frac{1}{p}}$$

$$> \frac{k_{\lambda}^{(1)}(\sigma)}{|a|^{1/q} b^{1/p}} \left[\int_{-\infty}^{\infty} \left(\frac{f(x)}{e^{\sigma ax}}\right)^{p} dx \right]^{\frac{1}{p}}, \qquad (5.66)$$

$$\left[\int_{-\infty}^{\infty} e^{-q\sigma ax} \left(\int_{e^{-by}}^{\infty} k_{\lambda}(e^{ax}, e^{-by}) g(y) dy \right)^{q} dx \right]^{\frac{1}{q}}$$

$$> \frac{k_{\lambda}^{(1)}(\sigma)}{|a|^{1/q} b^{1/p}} \left[\int_{-\infty}^{\infty} \left(\frac{g(y)}{e^{-\mu by}}\right)^{q} dy \right]^{\frac{1}{q}}, \qquad (5.67)$$

$$\int_{-\infty}^{\infty} g(y) \left(\int_{e^{-by}}^{\infty} k_{\lambda}(e^{ax}, e^{-by}) f(x) dx \right) dy$$

$$> \frac{k_{\lambda}^{(1)}(\sigma)}{|a|^{1/q} b^{1/p}} \left[\int_{-\infty}^{\infty} \left(\frac{f(x)}{e^{\sigma ax}}\right)^{p} dx \right]^{\frac{1}{p}} \left[\int_{-\infty}^{\infty} \left(\frac{g(y)}{e^{-\mu by}}\right)^{q} dy \right]^{\frac{1}{q}}. \qquad (5.68)$$

In particular, for

$$k_{\lambda}(u, 1) = \frac{|\ln u|^{\beta} (\min\{u, 1\})^{\alpha+\gamma}}{|u^{\alpha} - 1| (\max\{u, 1\})^{\lambda+\gamma}}$$

$(\beta, \alpha > 0, \sigma > -\alpha - \gamma)$, by Example 5.14, we have the following equivalent inequalities with

$$\frac{\Gamma(\beta+1)}{|a|^{1/q}b^{1/p}\alpha^{\beta+1}}\zeta\left(\beta+1,\frac{\sigma+\alpha+\gamma}{\alpha}\right)$$

being the best possible constant factor:

$$\left\{\int_{-\infty}^{\infty}e^{-p\mu by}\left[\int_{e^{-by}}^{\infty}\frac{|ax-by|^\beta(\min\{e^{ax},e^{-by}\})^{\alpha+\gamma}f(x)}{|e^{\alpha ax}-e^{-\alpha by}|(\max\{e^{ax},e^{-by}\})^{\lambda+\gamma}}dx\right]^p dy\right\}^{\frac{1}{p}}$$

$$> \frac{\Gamma(\beta+1)}{|a|^{1/q}b^{1/p}\alpha^{\beta+1}}\zeta\left(\beta+1,\frac{\sigma+\alpha+\gamma}{\alpha}\right)\left[\int_{-\infty}^{\infty}\left(\frac{f(x)}{e^{\sigma ax}}\right)^p dx\right]^{\frac{1}{p}}, \qquad (5.69)$$

$$\left\{\int_{-\infty}^{\infty}e^{q\sigma ax}\left[\int_{e^{-by}}^{\infty}\frac{|ax-by|^\beta(\min\{e^{ax},e^{-by}\})^{\alpha+\gamma}g(y)}{|e^{\alpha ax}-e^{-\alpha by}|(\max\{e^{ax},e^{-by}\})^{\lambda+\gamma}}dy\right]^q dx\right\}^{\frac{1}{q}}$$

$$> \frac{\Gamma(\beta+1)}{|a|^{1/q}b^{1/p}\alpha^{\beta+1}}\zeta\left(\beta+1,\frac{\sigma+\alpha+\gamma}{\alpha}\right)\left[\int_{-\infty}^{\infty}\left(\frac{g(y)}{e^{-\mu by}}\right)^q dy\right]^{\frac{1}{q}}, \qquad (5.70)$$

$$\int_{-\infty}^{\infty}g(y)\left[\int_{e^{-by}}^{\infty}\frac{|ax-by|^\beta \min\{e^{ax},e^{-by}\}}{|e^{ax}-e^{-by}|(\max\{e^{ax},e^{-by}\})^\lambda}f(x)dx\right]dy$$

$$> \frac{\Gamma(\beta+1)}{|a|^{1/q}b^{1/p}\alpha^{\beta+1}}\zeta\left(\beta+1,\frac{\sigma+\alpha+\gamma}{\alpha}\right)$$

$$\times \left[\int_{-\infty}^{\infty}\left(\frac{f(x)}{e^{\sigma ax}}\right)^p dx\right]^{\frac{1}{p}}\left[\int_{-\infty}^{\infty}\left(\frac{g(y)}{e^{-\mu by}}\right)^q dy\right]^{\frac{1}{q}}. \qquad (5.71)$$

Theorem 5.20 *Let M_2 be a positive constant, and $a \neq 0, b > 0$. If $k_2(\sigma) < \infty$, then the following statements (i), (ii) (iii) and (iv) are equivalent:*

(i) For any $f(x) \geq 0$, satisfying

$$0 < \int_{-\infty}^{\infty}\left(\frac{f(x)}{e^{\sigma ax}}\right)^p dx < \infty,$$

we have the following inequality:

$$\left[\int_{-\infty}^{\infty}e^{p\sigma_1 by}\left(\int_{e^{-by}}^{\infty}h(e^{ax+by})f(x)dx\right)^p dy\right]^{\frac{1}{p}}$$

$$> M_2\left[\int_{-\infty}^{\infty}\left(\frac{f(x)}{e^{\sigma ax}}\right)^p dx\right]^{\frac{1}{p}}. \qquad (5.72)$$

(ii) For any $g(y) \geq 0$, satisfying

$$0 < \int_{-\infty}^{\infty} \left(\frac{g(y)}{e^{\sigma_1 by}} \right)^q dy < \infty,$$

we have the following inequality:

$$\left[\int_{-\infty}^{\infty} e^{q\sigma ax} \left(\int_{e^{-by}}^{\infty} h(e^{ax+by}) g(y) dy \right)^q dx \right]^{\frac{1}{q}}$$

$$> M_2 \left[\int_{-\infty}^{\infty} \left(\frac{g(y)}{e^{\sigma_1 by}} \right)^q dy \right]^{\frac{1}{q}}. \tag{5.73}$$

(iii) For any $f(x) \geq 0$, satisfying

$$0 < \int_{-\infty}^{\infty} \left(\frac{f(x)}{e^{\sigma ax}} \right)^p dx < \infty,$$

and $g(y) \geq 0$, satisfying

$$0 < \int_{-\infty}^{\infty} \left(\frac{g(y)}{e^{\sigma_1 by}} \right)^q dy < \infty,$$

we have the following inequality:

$$\int_{-\infty}^{\infty} g(y) \left(\int_{e^{-by}}^{\infty} h(e^{ax+by}) f(x) dx \right) dy$$

$$> M_2 \left[\int_{-\infty}^{\infty} \left(\frac{f(x)}{e^{\sigma ax}} \right)^p dx \right]^{\frac{1}{p}} \left[\int_{-\infty}^{\infty} \left(\frac{g(y)}{e^{\sigma_1 by}} \right)^q dy \right]^{\frac{1}{q}}. \tag{5.74}$$

(iv)

$$\sigma_1 = \sigma \quad \text{and} \quad \frac{k_2(\sigma)}{|a|^{1/q} b^{1/p}} \geq M_2 \ (> 0).$$

If statement (iv) holds true, then the constant $M_2 = \frac{k_2(\sigma)}{|a|^{1/q} b^{1/p}}$ ($\in \mathbf{R}_+$) in (5.72), (5.73) and (5.74) is the best possible.

In particular, for $\sigma_1 = \sigma$,

$$h(u) = \frac{|\ln u|^{\beta} (\max\{u, 1\})^{\alpha}}{|u^{\lambda} - 1| (\min\{u, 1\})^{\alpha}}$$

$(\beta, \lambda > 0, \mu > \alpha)$, by Example 5.13, we have the following equivalent inequalities with

$$\frac{\Gamma(\beta+1)}{|a|^{1/q}b^{1/p}\lambda^{\beta+1}}\zeta\left(\beta+1,\frac{\mu-\alpha}{\lambda}\right)$$

being the best possible constant factor:

$$\left\{\int_{-\infty}^{\infty}e^{p\sigma by}\left[\int_{e^{-by}}^{\infty}\frac{|ax+by|^{\beta}(\min\{e^{ax+by},1\})^{\alpha}f(x)}{|e^{\lambda(ax+by)}-1|(\max\{e^{ax+by},1\})^{\alpha}}dx\right]^{p}dy\right\}^{\frac{1}{p}}$$

$$> \frac{\Gamma(\beta+1)}{|a|^{1/q}b^{1/p}\lambda^{\beta+1}}\zeta\left(\beta+1,\frac{\mu-\alpha}{\lambda}\right)\left[\int_{-\infty}^{\infty}\left(\frac{f(x)}{e^{\sigma ax}}\right)^{p}dx\right]^{\frac{1}{p}}, \tag{5.75}$$

$$\left\{\int_{-\infty}^{\infty}e^{q\mu ax}\left[\int_{e^{-by}}^{\infty}\frac{|ax+by|^{\beta}(\min\{e^{ax+by},1\})^{\alpha}g(y)}{|e^{\lambda(ax+by)}-1|(\max\{e^{ax+by},1\})^{\alpha}}dy\right]^{q}dx\right\}^{\frac{1}{q}}$$

$$> \frac{\Gamma(\beta+1)}{|a|^{1/q}b^{1/p}\lambda^{\beta+1}}\zeta\left(\beta+1,\frac{\mu-\alpha}{\lambda}\right)\left[\int_{-\infty}^{\infty}\left(\frac{f(x)}{e^{\sigma ax}}\right)^{p}dx\right]^{\frac{1}{p}}, \tag{5.76}$$

$$\int_{-\infty}^{\infty}g(y)\left[\int_{e^{-by}}^{\infty}\frac{|ax+by|^{\beta}(\min\{e^{ax+by},1\})^{\alpha}}{|e^{\lambda(ax+by)}-1|(\max\{e^{ax+by},1\})^{\alpha}}f(x)dx\right]dy$$

$$> \frac{\Gamma(\beta+1)}{|a|^{1/q}b^{1/p}\lambda^{\beta+1}}\zeta\left(\beta+1,\frac{\mu-\alpha}{\lambda}\right)$$

$$\times\left[\int_{-\infty}^{\infty}\left(\frac{f(x)}{e^{\sigma ax}}\right)^{p}dx\right]^{\frac{1}{p}}\left[\int_{-\infty}^{\infty}\left(\frac{g(y)}{e^{\sigma by}}\right)^{q}dy\right]^{\frac{1}{q}}. \tag{5.77}$$

Remark 5.21 In Theorem 5.20, if $\sigma_1 = \sigma$, $b < 0$, then replacing $-b$ by $b > 0$, we have the following equivalent inequalities with the best possible constant factor $\frac{k_2(\sigma)}{|a|^{1/q}b^{1/p}}$:

$$\left[\int_{-\infty}^{\infty}e^{-p\sigma by}\left(\int_{-\infty}^{e^{by}}h(e^{ax-by})f(x)dx\right)^{p}dy\right]^{\frac{1}{p}}$$

$$> \frac{k_2(\sigma)}{|a|^{1/q}b^{1/p}}\left[\int_{-\infty}^{\infty}\left(\frac{f(x)}{e^{\sigma ax}}\right)^{p}dx\right]^{\frac{1}{p}}, \tag{5.78}$$

$$\left[\int_{-\infty}^{\infty} e^{q\sigma ax}\left(\int_{-\infty}^{e^{by}} h(e^{ax-by})g(y)dy\right)^q dx\right]^{\frac{1}{q}}$$

$$> \frac{k_2(\sigma)}{|a|^{1/q}b^{1/p}}\left[\int_{-\infty}^{\infty}\left(\frac{f(x)}{e^{\sigma ax}}\right)^p dx\right]^{\frac{1}{p}},\tag{5.79}$$

$$\int_{-\infty}^{\infty} g(y)\left(\int_{-\infty}^{e^{by}} h(e^{ax-by})f(x)dx\right)dy$$

$$> \frac{k_2(\sigma)}{|a|^{1/q}b^{1/p}}\left[\int_{-\infty}^{\infty}\left(\frac{f(x)}{e^{\sigma ax}}\right)^p dx\right]^{\frac{1}{p}}\left[\int_{-\infty}^{\infty}\left(\frac{g(y)}{e^{-\sigma by}}\right)^q dy\right]^{\frac{1}{q}}.\tag{5.80}$$

In particular, for

$$h(u) = \frac{|\ln u|^\beta(\min\{u,1\})^{\alpha+\gamma}}{|u^\alpha - 1|(\max\{u,1\})^{\lambda+\gamma}}\quad(\beta,\alpha>0,\mu>-\alpha-\gamma),$$

by Example 5.14, we have the following equivalent inequalities with

$$\frac{\Gamma(\beta+1)}{|a|^{1/q}b^{1/p}\alpha^{\beta+1}}\zeta\left(\beta+1,\frac{\mu+\alpha+\gamma}{\alpha}\right)$$

being the best possible constant factor:

$$\left\{\int_{-\infty}^{\infty} e^{-p\sigma by}\left[\int_{-\infty}^{e^{by}}\frac{|ax-by|^\beta(\min\{e^{ax-by},1\})^{\alpha+\gamma}}{|e^{\alpha(ax-by)}-1|(\max\{e^{ax-by},1\})^{\lambda+\gamma}}f(x)dx\right]^p dy\right\}^{\frac{1}{p}}$$

$$> \frac{\Gamma(\beta+1)}{|a|^{1/q}b^{1/p}\alpha^{\beta+1}}\zeta\left(\beta+1,\frac{\mu+\alpha+\gamma}{\alpha}\right)\left[\int_{-\infty}^{\infty}\left(\frac{f(x)}{e^{\sigma ax}}\right)^p dx\right]^{\frac{1}{p}},\tag{5.81}$$

$$\left\{\int_{-\infty}^{\infty} e^{q\sigma ax}\left[\int_{-\infty}^{e^{by}}\frac{|ax-by|^\beta(\min\{e^{ax-by},1\})^{\alpha+\gamma}}{|e^{\alpha(ax-by)}-1|(\max\{e^{ax-by},1\})^{\lambda+\gamma}}g(y)dy\right]^q dx\right\}^{\frac{1}{q}}$$

$$> \frac{\Gamma(\beta+1)}{|a|^{1/q}b^{1/p}\alpha^{\beta+1}}\zeta\left(\beta+1,\frac{\mu+\alpha+\gamma}{\alpha}\right)\left[\int_{-\infty}^{\infty}\left(\frac{f(x)}{e^{\sigma ax}}\right)^p dx\right]^{\frac{1}{p}},\tag{5.82}$$

$$\int_{-\infty}^{\infty} g(y) \left[\int_{-\infty}^{e^{by}} \frac{|ax - by|^{\beta} (\min\{e^{ax-by}, 1\})^{\alpha+\gamma}}{|e^{\alpha(ax-by)} - 1|(\max\{e^{ax-by}, 1\})^{\lambda+\gamma}} f(x) dx \right] dy$$

$$> \frac{\Gamma(\beta+1)}{|a|^{1/q} b^{1/p} \alpha^{\beta+1}} \zeta\left(\beta+1, \frac{\mu+\alpha+\gamma}{\alpha}\right)$$

$$\times \left[\int_{-\infty}^{\infty} \left(\frac{f(x)}{e^{\sigma ax}}\right)^{p} dx \right]^{\frac{1}{p}} \left[\int_{-\infty}^{\infty} \left(\frac{g(y)}{e^{-\sigma by}}\right)^{q} dy \right]^{\frac{1}{q}}. \tag{5.83}$$

Corollary 5.22 *Let M_2 be a positive constant, and $a \neq 0, b > 0$. If*

$$k_{\lambda}^{(2)}(\sigma) = \int_{1}^{\infty} k_{\lambda}(u, 1) u^{\sigma-1} du > 0,$$

then the following statements (i), (ii) (iii) and (iv) are equivalent:
(i) For any $f(x) \geq 0$, satisfying

$$0 < \int_{-\infty}^{\infty} \left(\frac{f(x)}{e^{\sigma ax}}\right)^{p} dx < \infty,$$

we have the following inequality:

$$\left[\int_{-\infty}^{\infty} e^{p\mu_1 by} \left(\int_{e^{by}}^{\infty} k_{\lambda}(e^{ax}, e^{by}) f(x) dx \right)^{p} dy \right]^{\frac{1}{p}}$$

$$> M_2 \left[\int_{-\infty}^{\infty} \left(\frac{f(x)}{e^{\sigma ax}}\right)^{p} dx \right]^{\frac{1}{p}}. \tag{5.84}$$

(ii) For any $g(y) \geq 0$, satisfying

$$0 < \int_{-\infty}^{\infty} \left(\frac{g(y)}{e^{\mu_1 by}}\right)^{q} dy < \infty,$$

we have the following inequality:

$$\left[\int_{-\infty}^{\infty} e^{q\sigma ax} \left(\int_{e^{by}}^{\infty} k_{\lambda}(e^{ax}, e^{by}) g(y) dy \right)^{q} dx \right]^{\frac{1}{q}}$$

$$> M_2 \left[\int_{-\infty}^{\infty} \left(\frac{g(y)}{e^{\mu_1 by}}\right)^{q} dy \right]^{\frac{1}{q}}. \tag{5.85}$$

(iii) For any $f(x) \geq 0$, satisfying

$$0 < \int_{-\infty}^{\infty} \left(\frac{f(x)}{e^{\sigma ax}}\right)^{p} dx < \infty,$$

and $g(y) \geq 0$, *satisfying*

$$0 < \int_{-\infty}^{\infty} \left(\frac{g(y)}{e^{\mu_1 by}} \right)^q dy < \infty,$$

we have the following inequality:

$$\int_{-\infty}^{\infty} g(y) \left(\int_{e^{by}}^{\infty} k_\lambda(e^{ax}, e^{by}) f(x) dx \right) dy$$

$$> M_2 \left[\int_{-\infty}^{\infty} \left(\frac{f(x)}{e^{\sigma ax}} \right)^p dx \right]^{\frac{1}{p}} \left[\int_{-\infty}^{\infty} \left(\frac{g(y)}{e^{\mu_1 by}} \right)^q dy \right]^{\frac{1}{q}}. \tag{5.86}$$

(iii)

$$\mu_1 = \mu \ \text{and} \ \frac{k_\lambda^{(2)}(\sigma)}{|a|^{1/q} b^{1/p}} \geq M_2 \ (> 0).$$

If statement (iv) holds true, then the constant $M_2 = \frac{k_\lambda^{(2)}(\sigma)}{|a|^{1/q} b^{1/p}}$ *($\in \mathbf{R}_+$) in (5.84), (5.85) and (5.86) is the best possible.*

In particular, for $\mu_1 = \mu$,

$$h(u) = \frac{|\ln u|^\beta (\max\{u, 1\})^\alpha}{|u^\lambda - 1|(\min\{u, 1\})^\alpha}$$

($\beta, \lambda > 0, \mu > \alpha$), by Example 5.13, we have the following equivalent inequalities with

$$\frac{\Gamma(\beta + 1)}{|a|^{1/q} b^{1/p} \lambda^{\beta+1}} \zeta \left(\beta + 1, \frac{\mu - \alpha}{\lambda} \right)$$

being the best possible constant factor:

$$\left\{ \int_{-\infty}^{\infty} e^{p\sigma by} \left[\int_{e^{by}}^{\infty} \frac{|ax + by|^\beta (\min\{e^{ax}, e^{by}\})^\alpha f(x)}{|e^{\lambda ax} - e^{(\lambda+\alpha)by}|(\max\{e^{ax}, e^{by}\})^\alpha} dx \right]^p dy \right\}^{\frac{1}{p}}$$

$$> \frac{\Gamma(\beta + 1)}{|a|^{1/q} b^{1/p} \lambda^{\beta+1}} \zeta \left(\beta + 1, \frac{\mu - \alpha}{\lambda} \right) \left[\int_{-\infty}^{\infty} \left(\frac{f(x)}{e^{\sigma ax}} \right)^p dx \right]^{\frac{1}{p}}, \tag{5.87}$$

$$\left\{ \int_{-\infty}^{\infty} e^{q\mu ax} \left[\int_{e^{by}}^{\infty} \frac{|ax + by|^\beta (\min\{e^{ax}, e^{by}\})^\alpha g(y)}{|e^{\lambda ax} - e^{(\lambda+\alpha)by}|(\max\{e^{ax}, e^{by}\})^\alpha} dy \right]^q dx \right\}^{\frac{1}{q}}$$

$$> \frac{\Gamma(\beta + 1)}{|a|^{1/q} b^{1/p} \lambda^{\beta+1}} \zeta \left(\beta + 1, \frac{\mu - \alpha}{\lambda} \right) \left[\int_{-\infty}^{\infty} \left(\frac{g(y)}{e^{\mu by}} \right)^q dy \right]^{\frac{1}{q}}, \tag{5.88}$$

$$\int_{-\infty}^{\infty} g(y) \left[\int_{e^{by}}^{\infty} \frac{|ax+by|^{\beta}(\min\{e^{ax}, e^{by}\})^{\alpha}}{|e^{\lambda ax} - e^{(\lambda+\alpha)by}|(\max\{e^{ax}, e^{by}\})^{\alpha}} f(x)dx \right] dy$$

$$> \frac{\Gamma(\beta+1)}{|a|^{1/q} b^{1/p} \lambda^{\beta+1}} \zeta \left(\beta+1, \frac{\mu-\alpha}{\lambda} \right)$$

$$\times \left[\int_{-\infty}^{\infty} \left(\frac{f(x)}{e^{\sigma ax}} \right)^p dx \right]^{\frac{1}{p}} \left[\int_{-\infty}^{\infty} \left(\frac{g(y)}{e^{\mu by}} \right)^q dy \right]^{\frac{1}{q}}. \tag{5.89}$$

Remark 5.23 In Corollary 5.22, if $\mu_1 = \mu$, $b < 0$, then replacing $-b$ by $b > 0$, we have the following equivalent inequalities with the best possible constant factor $\frac{k_{\lambda}^{(2)}(\sigma)}{|a|^{1/q} b^{1/p}}$:

$$\left[\int_{-\infty}^{\infty} e^{-p\mu by} \left(\int_{-\infty}^{e^{-by}} k_{\lambda}(e^{ax}, e^{-by}) f(x)dx \right)^p dy \right]^{\frac{1}{p}}$$

$$> \frac{k_{\lambda}^{(2)}(\sigma)}{|a|^{1/q} b^{1/p}} \left[\int_{-\infty}^{\infty} \left(\frac{f(x)}{e^{\sigma ax}} \right)^p dx \right]^{\frac{1}{p}}, \tag{5.90}$$

$$\left[\int_{-\infty}^{\infty} e^{q\sigma ax} \left(\int_{-\infty}^{e^{-by}} k_{\lambda}(e^{ax}, e^{-by}) g(y)dy \right)^q dx \right]^{\frac{1}{q}}$$

$$> \frac{k_{\lambda}^{(2)}(\sigma)}{|a|^{1/q} b^{1/p}} \left[\int_{-\infty}^{\infty} \left(\frac{g(y)}{e^{-\mu by}} \right)^q dy \right]^{\frac{1}{q}}, \tag{5.91}$$

$$\int_{-\infty}^{\infty} g(y) \left(\int_{-\infty}^{e^{-by}} k_{\lambda}(e^{ax}, e^{-by}) f(x)dx \right) dy$$

$$> \frac{k_{\lambda}^{(2)}(\sigma)}{|a|^{1/q} b^{1/p}} \left[\int_{-\infty}^{\infty} \left(\frac{f(x)}{e^{\sigma ax}} \right)^p dx \right]^{\frac{1}{p}} \left[\int_{-\infty}^{\infty} \left(\frac{g(y)}{e^{-\mu by}} \right)^q dy \right]^{\frac{1}{q}}. \tag{5.92}$$

In particular, for

$$k_{\lambda}(u,1) = \frac{|\ln u|^{\beta}(\min\{u, 1\})^{\alpha+\gamma}}{|u^{\alpha} - 1|(\max\{u, 1\})^{\lambda+\gamma}}$$

$(\beta, \alpha > 0, \mu > -\alpha - \gamma)$, by Example 5.14, we have the following equivalent inequalities with

$$\frac{\Gamma(\beta+1)}{|a|^{1/q} b^{1/p} \alpha^{\beta+1}} \zeta \left(\beta+1, \frac{\mu+\alpha+\gamma}{\alpha} \right)$$

being the best possible constant factor:

$$\left\{ \int_{-\infty}^{\infty} e^{-p\mu by} \left[\int_{-\infty}^{e^{-by}} \frac{|ax - by|^{\beta} (\min\{e^{ax}, e^{-by}\})^{\alpha+\gamma} f(x)}{|e^{\alpha ax} - e^{-\alpha by}|(\max\{e^{ax}, e^{-by}\})^{\lambda+\gamma}} dx \right]^{p} dy \right\}^{\frac{1}{p}}$$

$$> \frac{\Gamma(\beta+1)}{|a|^{1/q} b^{1/p} \alpha^{\beta+1}} \zeta \left(\beta+1, \frac{\mu+\alpha+\gamma}{\alpha} \right) \left[\int_{-\infty}^{\infty} \left(\frac{f(x)}{e^{\sigma ax}} \right)^{p} dx \right]^{\frac{1}{p}}, \qquad (5.93)$$

$$\left\{ \int_{-\infty}^{\infty} e^{q\sigma ax} \left[\int_{-\infty}^{e^{-by}} \frac{|ax - by|^{\beta} (\min\{e^{ax}, e^{-by}\})^{\alpha+\gamma} g(y)}{|e^{\alpha ax} - e^{-\alpha by}|(\max\{e^{ax}, e^{-by}\})^{\lambda+\gamma}} dy \right]^{q} dx \right\}^{\frac{1}{q}}$$

$$> \frac{\Gamma(\beta+1)}{|a|^{1/q} b^{1/p} \alpha^{\beta+1}} \zeta \left(\beta+1, \frac{\mu+\alpha+\gamma}{\alpha} \right) \left[\int_{-\infty}^{\infty} \left(\frac{g(y)}{e^{-\mu by}} \right)^{q} dy \right]^{\frac{1}{q}}, \qquad (5.94)$$

$$\int_{-\infty}^{\infty} g(y) \left[\int_{-\infty}^{e^{-by}} \frac{|ax - by|^{\beta} (\min\{e^{ax}, e^{-by}\})^{\alpha+\gamma} f(x)}{|e^{\alpha ax} - e^{-\alpha by}|(\max\{e^{ax}, e^{-by}\})^{\lambda+\gamma}} dx \right] dy$$

$$> \frac{\Gamma(\beta+1)}{|a|^{1/q} b^{1/p} \alpha^{\beta+1}} \zeta \left(\beta+1, \frac{\mu+\alpha+\gamma}{\alpha} \right)$$

$$\times \left[\int_{-\infty}^{\infty} \left(\frac{f(x)}{e^{\sigma ax}} \right)^{p} dx \right]^{\frac{1}{p}} \left[\int_{-\infty}^{\infty} \left(\frac{g(y)}{e^{-\mu by}} \right)^{q} dy \right]^{\frac{1}{q}}. \qquad (5.95)$$

Acknowledgements B. Yang: This work is supported by the National Natural Science Foundation (No. 61772140), and Science and Technology Planning Project Item of Guangzhou City (No. 201707010229). M. Th. Rassias: This work was carried out under the generous research support of the University of Zurich and the John S. Latsis Foundation, spending also research time at the vibrant environment of the Institute for Advanced Study, Princeton.

References

1. Kuang, J.C.: Real and Functional Analysis (continuation) (sec. vol.). Higher Education Press, Beijing, China (2015)
2. Kuang, J.C.: Applied Inequalities. Shangdong Science and Technology Press, Jinan, China (2004)
3. Wang, Z.X., Guo, D.R.: Introduction to Special Functions. Science Press, Beijing (1979)

Bibliography

4. Weyl, H.: Singulare Integral Gleichungen mit Besonderer Berucksichtgung des Fourierschen Integral Theorem. Inaugural-Dissertation, Gottingen (1908)
5. Pan, Y.L., Wang, H.T., Wang, F.T.: On Complex Functions. Science Press, Beijing, China (2006)
6. Zhong, Y.Q.: On Complex Functions. Higher Education Press, Beijing, China (2004)

7. Edwards, H.M.: Riemann's Zeta Function. Dover Publications, New York (1974)
8. Zhao, D.J.: On a refinement of Hilbert double series theorem. Math. Pract. Theory **23**(1), 85–90 (1993)
9. Gao, M.Z.: On an improvement of Hilbert's inequality extended by Hardy-Riesz. J. Math. Res. Expos. **14**(2), 255–259 (1994)
10. Gao, M.Z., Yang, B.C.: On the extended Hilbert's inequality. Proc. Am. Math. Soc. **126**(3), 751–759 (1998)
11. Yang, B.C., Debnath, L.: On new strengthened Hardy-Hilbert's inequality. Internat. J. Math. Math. Soc. **21**(2), 403–408 (1998)

Printed in the United States
By Bookmasters